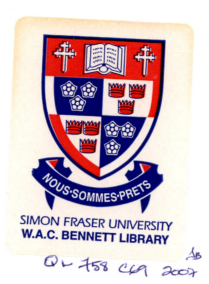

Predator–Prey Dynamics

The Role of Olfaction

Predator–Prey Dynamics

The Role of Olfaction

Michael R. Conover

CRC Press
Taylor & Francis Group
Boca Raton   London   New York

CRC Press is an imprint of the
Taylor & Francis Group, an informa business

CRC Press
Taylor & Francis Group
6000 Broken Sound Parkway NW, Suite 300
Boca Raton, FL 33487-2742

Library of Congress Cataloging-in-Publication Data

Conover, Michael R.
   Predator-prey dynamics : role of olfaction / Michael Conover.
      p. cm.
   Includes bibliographical references.
   ISBN-13: 978-0-8493-9270-2 (alk. paper)
   ISBN-10: 0-8493-9270-5 (alk. paper)
   1. Predatory animals--Sense organs. 2. Smell. 3. Predation (Biology) I. Title.

QL758.C69 2007
591.5'3--dc22   2006032886

Visit the Taylor & Francis Web site at
http://www.taylorandfrancis.com

and the CRC Press Web site at
http://www.crcpress.com

Dedication

Dedicated to
Robert and Phyllis Conover and
Robert and Rosalie O'Neal.

Contents

Preface

We know a great deal about how animals hide from predators and marvel at how accomplished they are at hiding. We are fascinated with their camouflage coloration and how some insects hide by mimicking leaves, flowers, twigs, thorns, bark, or hundreds of other things. We are amazed at how even a large deer can disappear into its background when it freezes into position.

In all of these cases, animals are hiding from predators that use their vision to locate them. Yet, many predators rely on olfaction to locate prey. This raises the question of whether animals are capable of hiding from predators that are using olfaction and, if so, how animals accomplish this feat. The realization that so little is known about how animals hide from olfactory predators led me to investigate this topic and ultimately to write this book.

I begin by explaining how this book is organized. The first five chapters examine the mechanics of olfaction and how predators detect and locate prey using odor trails left behind as animals move across the ground or by using airborne odor plumes that animals create constantly. Chapters 6 through 8 explain the physics of airflow and examine where turbulence and updrafts are likely to occur. It is in these places where the prey of olfactory predators can successfully hide. Chapter 9 discusses the difficulties that animals face when they have to hide from both olfactory and visual predators while completing all of the other tasks that are required if they are to survive and reproduce. Chapters 10 through 13 examine the question of whether mammals and birds actually hide in areas where turbulence and updrafts are prevalent and if doing so helps them avoid death from olfactory predators. In Chapter 14, I redefine some ecological terms based on the physics of airflow. Finally, in Chapter 15, I summarize the theory that there is a constant struggle between olfactory predators and animals that seek to hide from them, and I discuss implications of this theory.

I hope you enjoy reading this book and gain a new perception of our environment. The olfactory predator's world is different from the one that we perceive as humans. In our world, lines are straight because light travels in straight lines. In the world of olfactory predators, lines bend because odors flow around physical objects. So, welcome to a new world—the world of olfactory predators and their prey.

Acknowledgments

This book would not have been possible without the help of Rosanna Vail. She spent tireless hours checking references, editing, and trying to make sense of my awkward style of writing. I acknowledge and thank her for her great skill, good humor, and willingness to redo things countless times because of my inability to explain what I needed the first time.

I thank Holly Archibald, Denise Conover, Phyllis Conover, and Philip Parisi for their help editing former drafts of this book. I am grateful to the many scientific journals and publishing companies that allowed me to use without charge figures that originally appeared in their publications. I thank numerous photographers who allowed me to include their photos in my book. These photographers and the original source of figures are mentioned in the figure legends. Rosanna Vail drew all of the figures and drawings that appear in this book. I took the photos that lack photo credits.

chapter one

Olfactory predators and odorants

A primary defense against predators that is used by most mammals and birds is to try to avoid detection. Many predators are visual predators and rely on light reflected off their prey to locate the prey. Evolution has provided these predators with pronounced visual acuity to see their prey at great distances and to detect even tiny movements. To survive, prey have evolved numerous methods to make it harder for visual predators to see them. Many have camouflaged or disruptive coloration and patterns that allow them to blend into their backgrounds. When threatened, they remain motionless or have other behavioral patterns that reinforce their cryptic nature. For instance, when Canada geese (*Branta canadensis*) are hiding, they keep their long necks and heads close to the ground rather than upright so that predators cannot detect them by their unique silhouette (Latin names for a species are given the first time a species is mentioned in each chapter; they also are provided in Appendix 1).

Humans are visual predators; we use vision to hunt for our food. Because we rely on vision, we are familiar with the various methods that mammals and birds employ to avoid being seen, and we are aware of the constant struggle between prey that try to hide and predators that try to locate them using visual cues. Although we are usually oblivious to it because our own sense of olfaction is so poorly developed, an equally dynamic struggle occurs between predators that use olfaction to locate prey and the animals that seek to avoid detection through olfaction.

All animals constantly release a stream of chemicals into the environment. Many of these chemicals are by-products of metabolism. Many predators have evolved the ability to detect these chemicals and use them to locate the animal releasing them. The ability to detect the presence of chemicals in the air is called *olfaction,* and predators that primarily use this modality to locate prey are called *olfactory predators*. Most chemicals cannot be detected by olfactory systems, but those that can are called *odorants*. Although it is impossible for animals to stop releasing odorants, they can use atmospheric conditions and environmental variables to make it difficult for olfactory predators to find them. This book explores the struggle for survival between olfactory predators and their prey.

Finding prey requires two distinctive skills. The first is to detect the presence of prey, and the second is to find its exact location. Olfactory predators use this modality to accomplish both tasks. Once they have used olfactory cues to get within a few meters of the prey, olfactory predators may shift to another modality, often vision, to pinpoint the prey's exact location, but I still consider these animals olfactory predators because olfaction is the primary modality they use to detect and locate prey. Olfactory predators come from the ranks of almost all taxa, and among terrestrial vertebrates, olfactory predators include some amphibians, reptiles, and mammals. Yet, as far as numbers are concerned, most are marine invertebrates and fish. For simplicity, I have limited this book to a study of terrestrial vertebrates, but the information contained here will still be useful to marine biologists because the principles of airflow and water flow are similar. Hence, there are great similarities in how olfactory predators hunt and how their prey use turbulence to hide from them regardless of whether those predators are terrestrial or aquatic.

Before I discuss how predators use olfaction to locate prey, I first need to explain how olfactory organs function, why only some chemicals have a detectable odor, what causes odorants either to persist on a surface or to evaporate, and what causes odorants to move through the atmosphere.

Olfactory organs of vertebrates

With the notable exception of humans and some other primates, mammals have two major chemosensory organs, both located in the nasal cavities. These are the main olfactory system (MOS) and the vomeronasal (VNO) system, each of which has its own neural pathway and serves different functions. The MOS is designed to sample the air for odorants that may convey information about the presence and location of food or predators. Odorants are delivered to the MOS through the normal process of inhaling. In mammals, the VNO functions primarily to sample for species-specific pheromones involved in endocrine responses and reproductive behaviors. Mammals deliver odorants contained in liquids, such as urine, directly to the VNO through a behavior called *flehmen*, during which the tip of the tongue is placed over the VNO duct in the mouth while sniffing. In snakes, both the VNO and MOS are used for prey detection. Snakes deliver odorants to the VNO by first flicking their tongue out of their mouth to collect odorants and then touching their tongue to the opening of their VNO. Given our focus on how predators, especially mammalian predators, locate prey through olfaction, we do not deal further with the VNO, and all future comments pertain to the MOS.

Vertebrate nasal cavities are complex aerodynamic structures that mammals use to sample the external environment for odorants by inhaling. In humans, the odorant receptors occupy a 1-cm^2 area at the uppermost arched roof of the nasal cavity about 7 cm from the external openings to the nose (symbols used in this book appear in Appendix 2). The receptors are in an olfactory slit about 1 mm wide but several millimeters in height and length. Hence, the olfactory slit is like a long, narrow passageway with high walls. Only 5 to 10% of an inhaled breath actually passes through this olfactory slit, and this proportion does not seem to change with increasing airflow (Mozell et al. 1991). It is in the olfactory slit where millions of odorant receptor cells are matted together to form the olfactory epithelium.

Olfactory receptor cells vary little across the diverse array of terrestrial species because they all have a common mission—to capture and bind chemicals suspended in the air and to transmit information about their presence into neural signals. Olfactory receptor cells are sensory neurons that connect the brain to the atmosphere. It is through them that the brain gains information about environmental odorants. One end of these cells is in contact with the air and has a knob from which 10 to 30 cilia project (Moran et al. 1991; Farbman 2000). It is on these cilia that the odorant receptors are located (Figure 1.1). Each odorant receptor reacts to a single odorant or single group of odorants that are chemically similar.

Mammals, including humans, have about 1000 different odorant receptors, each of which is a protein that is produced by a specific gene (Nef 1993; Axel 1995). One indication of the importance of olfaction to mammals is that the 1000 genes used to produce these different odorant receptors constitute about 3% of an individual's genes.

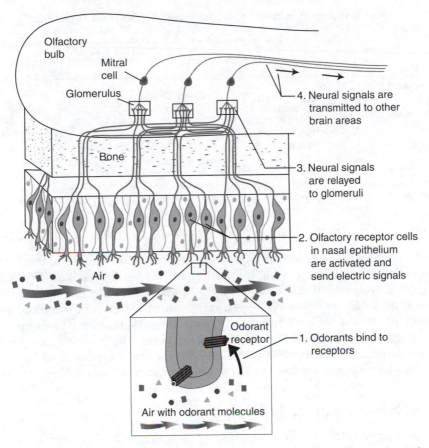

Figure 1.1 The olfactory system of mammals. (Adapted from www.nobelprize.org and used with permission of the Nobel Foundation.)

When an odorant receptor is activated by the presence of the odorant to which it is sensitive, the odorant receptor releases an electrical signal down its axon. These axons are connected to glomeruli located in the olfactory bulb (Figure 1.1); the glomeruli act as relay circuits for neural messages. There are about 2000 glomeruli, about twice as many as there are distinctive odorant receptors. Each glomerulus only receives axons from one specific type of odorant receptor. Humans have a total of 5 to 10 million receptor cells, so each of the 1000 different odorant receptors is repeated several thousand times in a person's olfactory epithelium. Hence, each glomerulus is connected to several thousand receptor cells via their axons, each with the same type of odorant receptor.

Each glomerulus in turn is connected to a mitral nerve cell, and each mitral cell is activated by a single glomerulus (Figure 1.1). The long axons of the mitral cells relay the olfactory information to specific parts of the brain's olfactory cortex. It is here that information from several types of odorant receptors is combined into a pattern that characterizes an odor. Although humans have only 1,000 odorant receptors, they can distinguish and form memories for more than 10,000 different odorants. The brain's olfactory cortex can do this because each odorant activates a unique combination of odorants (Axel 1995). For example, nonanoic acid activates odorant receptors 1, 4, 5, 7, 8, 10, 12, and 14 and has a nut-like smell (Table 1.1); a similar chemical, nonanol, activates receptors 4, 5, 7, 10, and 12 and has a floral smell.

Table 1.1 Odorant Receptors Activated by Similar Odorants and How Their Order Is Perceived by Humans

Odorants	Odor's description	Odorant receptors													
		1	2	3	4	5	6	7	8	9	10	11	12	13	14
Hexanoic acid	Rancid, sour	—	—	—	—	5	—	—	—	—	—	—	—	—	—
Hexanol	Sweet, herbal	—	2	—	—	—	6	—	—	—	—	—	—	—	—
Octanoic acid	Rancid, repulsive	1	—	—	4	5	—	7	8	—	10	11	12	—	—
Octanol	Sweet, orange	—	—	—	4	5	—	7	—	—	10	—	—	—	—
Nonanoic acid	Nut-like, cheese	1	—	—	4	5	—	7	8	—	10	—	12	—	14
Nonanol	Rose, floral	—	—	—	4	5	—	7	—	—	10	—	12	—	—

Source: Adapted from Howard Hughes Medical Institute, *Serving Science: '05 Annual Report*, Chevy Chase, MD, 2005, and www.nobelprize.org and used with permission of the Nobel Prize Foundation.

Comparing the olfactory ability of humans to other mammals

The olfactory system of humans and the neural processes that underlie them are similar to those of other mammals, but what differs is acuity (Passe and Walker 1985; Walker and Jennings 1991). Domestic dogs (*Canis familiaris*) have more olfactory receptor cells than do humans, and their receptor cells respond to much lower concentrations of specific odorants than receptor cells of humans (Wright 1964; Walker and Jennings 1991). For example, dogs are 300 times more sensitive than humans to amyl acetate (Krestel et al. 1984), 1,000 to 10,000 times more sensitive to alpha-ionone (Marshall and Moulton 1981), and 1 million times more sensitive to acetic acid and butyric acid (Wright 1964).

Use of olfaction by birds to locate food

A highly developed olfactory system would not provide much help finding food for birds that forage high in plant canopies or catch flying insects. Hence, it is not surprising that the olfactory ability of most birds, especially passerines, is poorer than that of mammals. Yet, a few avian species have a keen sense of smell and use it to locate prey. Hummingbirds (Trochilidae), honeyguides (Indicatoridae), and kiwis (*Apteryx* spp.) use olfaction to locate and discriminate among foods (Mason and Clark 2000). Olfactory ability is particularly useful for birds that forage on carcasses. Turkey vultures (*Cathartes aura*) are attracted by the smell of ethyl mercaptan, a by-product of decomposition, and can locate decomposing carcasses in the absence of visual cues (Stager 1967; Mason and Clark 2000). Still, in all of these birds, vision is more important than olfaction in locating food, especially when trying to hone in on it over long distances. For this reason, almost all birds rely on their eyesight to locate food and are visual predators; many mammals, especially those that hunt at night, are olfactory predators (Caro 2005). For instance, most birds that depredate nests find the nests using visual cues, but mammals find them using olfaction (Rangen et al. 2000; Caro 2005).

There is one bird group that has members that can qualify as olfactory predators. That group is the Procellariiformes or tube-nose sea birds (Figure 1.2). They feed at sea on fish, floating refuse and fish carcasses, and they vary in size from the tiny 50-g Wilson's storm-petrel (*Oceanites oceanicus*) to the wandering albatross (*Diomedea exulans*) with its 3-m wingspan. The olfactory systems of Procellariiformes are highly developed and are capable of distinguishing faint odorants. For example, Wilson's storm-petrels, black-bellied storm-petrels (*Fregetta tropica*), white-chinned petrels (*Procellaria aequinoctialis*), cape petrels (*Daption capensis*), giant-petrels (*Macronectes* spp.), and greater shearwaters (*Puffinus gravis*)

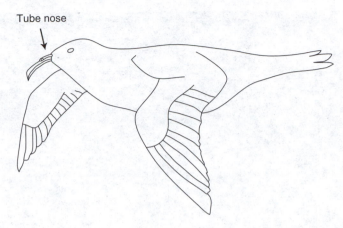

Tube nose

Figure 1.2 The members of the avian order Procellariiformes are unique among birds in having external tubular nostrils, which can be seen on this drawing of a petrel. They use their olfactory ability to locate food at sea.

are attracted from great distances by the smell of cod liver oil (Grubb 1972; Lequette et al. 1989).

Antarctic krill (*Euphausia superba*) is the key component for the Antarctic food chain. Krill live in schools that sometimes cover thousands of cubic meters but are widely distributed across the vast Antarctic Ocean. Regardless of whether a sea bird feeds directly on krill or on the animals that feed on them, sea birds must locate krill schools on a regular basis to survive. Fortunately, there are olfactory cues that procellariiform sea birds can use to help locate these schools. When krill are attacked and chewed up by predators such as fish, they release odorants. In one experiment involving the creation of oil slicks at sea, vegetable oil slicks containing krill odorants attracted five times as many procellariiform sea birds as vegetable oil slicks lacking the odorants (Nevitt 1999). Furthermore, when krill eat phytoplankton, the latter releases dimethyl sulfide, and Nevitt (1999) hypothesized that procellariiform sea birds may also use this odorant to locate schools of krill. To test his hypothesis, he compared the number of birds attracted to vegetable oil slicks containing dimethyl sulfide to the number attracted to plain vegetable oil slicks. He found that some species of procellariiform sea birds, including white-chinned petrels and two species of storm-petrels, were attracted by the dimethyl sulfide, but that albatrosses and cape petrels were not.

Although all of these birds use olfaction to locate rich hunting grounds, they have to rely on their vision to pluck the food from the water. So, both modalities are essential for their survival.

In the vastness of the oceans, it also can be hard for a sea bird to find the little island where its nest is located, and once that is accomplished, it may be difficult for the sea bird to find its own burrow from among the thousands that might be on the same island. Some procellariiform sea birds use olfaction to accomplish these tasks. Leach's storm-petrels (*Oceanodroma leucorrhoa*) and wedged-tailed shearwaters (*Puffinus pacificus*) use olfaction first to help locate their colonies and then to find their nesting burrows (Grubb 1973, 1974; Shallenberger 1975).

Which modality is most important to snakes in locating prey?

Most snakes are ambush hunters and wait for their prey to approach them. They usually have little need to locate prey or to track their movements at great distances. The reason

Figure 1.3 Rattlesnakes use infrared radiation to detect prey. (Courtesy of C.R. Madsen.)

for this is that most snakes travel too slowly to be able to close the distance with their quarry if they were tracking it. However, there are exceptions to this, and these usually involve prey that move even slower than snakes. One such prey item is a slug, and garter snakes (*Thamnophis* spp.) in California that specialize in eating banana slugs (*Ariolimax* spp.) locate their quarry using olfaction to identify and follow the trails left by passing slugs (Alcock 2001). Earthworms and snails are also slow moving, and the brown snake (*Storeria dekayi*) relies on olfaction to trail and locate them (Noble and Clausen 1936). Many poisonous snakes inject venom into their victims and then release the victims. The victims usually run several meters before death or paralysis stops them. The snakes then use their olfactory abilities to follow the path taken by their victims. Hence, olfaction is more important to snakes than vision or hearing during those occasions when they have to track their quarry, be they slugs or envenomed mammals.

As ambush hunters, snakes have to locate and identify any object that travels within their striking distance and determine if it is a suitable prey item and should be attacked. Snakes employ both vision and olfaction in prey location and identification, and some snakes, such as pit vipers, detect prey using infrared radiation (Figure 1.3). Often, both vision and olfactory cues are necessary for prey identification. For example, some snakes will not strike at live animals that are odorless or immobile (Burghardt 1966; Czaplicki and Porter 1974; Caro 2005) but the latter are still important. Northern water snakes (*Nerodia sipedon*) rarely hunt in water bodies lacking the odor of fish but do so once fish odor has been added (Drummond 1979). Once hunting underwater, these snakes attack moving objects more frequently than stationary ones, but they also attack pebbles that have been rubbed against fish but will not attack odorless pebbles (Drummond 1979).

For terrestrial snakes, visual cues are less important than olfactory ones (Czaplicki and Porter 1974). Burghardt (1966) found that snakes ignored prey that lacked olfactory cues, but that snakes that had lost their eyesight continued to attack and eat prey. Newborn garter snakes will even strike at cotton swabs daubed with the odor of prey but ignored odorless swabs (Burghardt and Hess 1968; Alcock 2001).

Snakes are a serious threat to eggs and the young of birds (Figure 1.4). Yet, most snakes will not eat eggs located in artificial nests (Davidson and Bollinger 2000; Marini and Melo 1998). Apparently, eggs in artificial nests do not have the same scent as those in real nests, and many snakes do not recognize them as food. Support for this comes from the research of Marini and Melo (1998), who gave quail eggs maintained at room temperature to 86 snakes of 22 species and eggs warmed to normal incubation temperatures to 17 individuals

Figure 1.4 This snake was swallowing a gosling head first when it was disturbed by the photographer. By the time this photograph was taken, the snake had already regurgitated the chick and was beginning its escape. The gosling's body is in the upper right corner of the photo, its head is in the center of the photo and covered with a bloody mucous, and the snake is in the lower left corner.

of 9 snake species. All species tested were known to depredate bird nests in the field, yet every snake refused to eat an egg. Hence, olfactory cues are important for many snakes in helping them identify what is potential food and what is not.

Which modality is most important to predatory mammals in locating prey?

As discussed in this book, there is no easy answer to the question of which modality is most important to predatory mammals in locating prey because it always depends on the circumstances that exist at the time. If an animal is hiding and the predator is fortunate enough to be downwind of it, then olfaction may be the more important modality. If the prey animal is moving and light conditions are good, then vision may be more important. The importance of a modality also changes as the hunt progresses. A mammalian predator may use olfaction first to detect the presence of an animal and then to track the quarry to within a few meters. It may then freeze and visually locate its quarry before pouncing and making the kill (Figure 1.5).

Figure 1.5 Mammalian predators often use their vision when close enough to their quarry to kill it. (Courtesy of Guy Connolly and U.S. Department of Agriculture's Wildlife Services.)

Figure 1.6 American badgers locate food primarily by olfaction. (Courtesy of U.S. Department of Agriculture's Wildlife Services.)

The relative importance of vision and olfaction in locating food also varies among mammalian species. Striped skunks (*Mephitis mephitis*) find prey mainly by smell and sound (Nams 1997); feral hogs (*Sus scrofa*) and American badgers (*Taxidea taxus*; Figure 1.6) use mostly olfaction (Green and Anthony 1989). Canids are more likely to use olfaction to find food than felids. Brown hyenas (*Hyaena brunnea*) use olfaction to locate food, but spotted hyenas (*Crocuta crocuta*) use both olfaction and vision to find food (Mills 1990). A likely explanation for this difference between hyena species is that brown hyenas feed mainly by scavenging carcasses and eating fruit, and spotted hyenas feed mainly by killing large mammals. Mills (1990) observed spotted hyenas in open habitat and judged that they used their vision first to detect potential prey in 76 of 126 (60%) contacts.

Mech (1970) noted that wolves (*Canis lupis*) use three different methods to locate prey: vision to spot prey, olfaction to follow depositional odor trails, and olfaction to scent the animal directly. Of these methods, Mech (1970) believed that vision was the least important for wolves and was rarely used to locate prey (moose, *Alces alces*) on Isle Royale National Park located on Lake Superior. He observed that wolves could detect a moose when within 300 m downwind of it. However, when the moose was downwind or crosswind to them, the moose often escaped the wolves' attention. Mech found that hunting episodes by wolves usually began when they directly scented moose rather than when locating them by following their depositional odor trails. In more open areas, such as those that occur on Mount McKinley in Alaska, vision plays a greater role in prey detection by wolves (Murie 1944).

Characteristics of odorants

To be detected by a terrestrial mammal through olfaction, a chemical must be capable of suspension in the air. Hence, most odorants are small molecules with a molecular weight less than 300 and a vapor pressure above 0.01 mmHg at ambient temperatures (Rawson 2000). DeVos et al. (1990) identified 529 chemicals that could be detected by humans. Of these, 98% were organic compounds. The nonorganic odorants include ammonia (NH_3), arsine (AsH_3), bromine (Br_2), hydrogen chloride (HCl), hydrogen sulfide (H_2S), nitrogen dioxide (NO_2), ozone (O_3), sulfur dichloride (SCl_2), and sulfur dioxide (SO_2). Most of the organic compounds contain fewer than 10 carbon atoms, and none contain more than 20.

Humans use their olfactory systems to detect odorants at the height of their nose, 1.5 to 2 m above the ground. Hence, human olfactory systems have evolved to detect volatile

Figure 1.7 By placing their noses close to the ground, feral hogs can inhale heavy organic chemicals that are lying on the ground. (Courtesy of U.S. Department of Agriculture's Wildlife Services.)

odorants that would normally occur at such heights above the ground. Compared to humans, however, tracking predators such as a feral pig or a bloodhound have probably evolved the ability to detect organic molecules of much heavier molecular weight and lower volatility (Figure 1.7). By placing their noses close to the ground and inhaling rapidly, bloodhounds can blow these nonvolatile chemicals off surfaces and then inhale them so that they will reach the mucous membrane covering the olfactory cilia. Once these molecules are there, the animal may have special proteins that help shuttle these large organic chemicals across the mucous membrane to the cilia. The ability to detect the presence of heavy organic chemicals would be beneficial for tracking predators because these chemicals are more likely to make up the path trail than the more volatile odorants.

Perception of odor mixtures

Most objects simultaneously give off several odorants rather than a single one, and it is this mixture of odorants that is detected by an animal. Mammalian olfactory systems have evolved the ability to detect this mixture of odorants and identify the odorants as a specific odor, although exactly how olfactory systems accomplish this task is unknown. That is, olfaction is similar to vision because it is a synthetic sense. For instance, the aroma of coffee contains over 800 different chemicals, at least 30 of which are required to be present to produce coffee's characteristic smell. This phenomenon of identifying a mixture of odorants as a single, unique odor is further complicated because when humans smell coffee, they cannot identify the individual odorants that make up its smell, but they can identify the odorants if each one is presented by itself. What happens is that the presence of these individual odorants disappears into the odor of coffee similar to the way the colors blue and yellow disappear into the color green (Laing 1991). Because olfaction is a synthetic sense, it is difficult to predict how a mixture of two different odorants will be perceived by the olfactory system. Sometimes, the combination of two odorants creates an odor stronger than when either odorant is presented by itself, and sometimes the combination creates a weaker odor. Sometimes, one odorant completely masks the presence of another. At other times, two odorants create a novel third odor.

Sources of odorants from mammals and birds

Birds and mammals are rich sources of odorants. Some odorants are metabolic in origin and are exhaled with each breath. These include carbon dioxide, acetone, and several phenolic compounds. Many blood-sucking insects, such as mosquitoes and tsetse flies (*Glossina* spp.), use these exhaled odorants to locate mammals (Voskamp et al. 1998).

Mammals have skin glands that serve many purposes and secrete a wide variety of substances. In some cases, the function of these glands is to produce specialized scents, or *pheromones*, that are used to communicate with conspecifics (that is, individuals of the same species). Secretions from salivary glands, eye glands, and anal sacs all produce odorants that are used to communicate with conspecifics. Odorants from the anal sac gland along with those in urine and feces are commonly used in scent marking. These odorants inform other conspecifics about the animal's sex, age, health, position in the social hierarchy, reproductive condition, and territorial status. Scent marking creates a record of the actions and locations of individuals and can notify other conspecifics about pack membership and territorial boundaries (Barrette and Messier 1980; Singer et al. 1997). These odorants can also convey information about the degree of genetic relatedness between the animal inhaling the odorants and the animal that produced them (Tegt 2004). These pheromones are usually not single odorants but complex mixtures of them (Stoddard 1980), and although their purpose is to communicate with conspecifics, the same chemical cues are available to predators trying to locate prey.

Most skin glands produce odorants but do so only as a by-product of their normal function. Eccrine sweat glands (true sweat glands) are widely distributed on the skin of humans and some primates, where they help regulate body temperature. In most mammals, however, they are located mainly on foot pads, where their function is to keep the pads pliable, allowing firm traction between the foot and the ground so the animal will not slip when accelerating, stopping, or turning (Aldeman et al. 1975). Sweat glands produce many odorants. Human axillary sweat, for instance, contains up to 300 different odorants (Sommerville and Gee 1984 as cited by Doving 1990).

Most birds and mammals have soft, pliable, waterproof surfaces composed of fur, feathers, or skin. These characteristics are achieved by keeping the surface clean and covered by a thin film of nonpolar chemicals (for example, lipids and oils). In mammals, these chemicals are secreted by glands distributed across the entire surface of the animal (for example, sweat glands in humans). Birds have fewer skin glands than mammals and no eccrine sweat glands. They do possess an uropygial gland that secretes a fluid containing fatty acids, lipids, and wax. When preening, birds squeeze the oil out of this gland with their bills and then spread the oil through their feathers.

Most mammals spend a great deal of time grooming their fur and feet. In doing so, they coat themselves with saliva, which is itself a rich source of odorants. Proteins found in saliva include lipases, amylases, ribonucleases, peroxidases, proteinases, peptidases, glycoproteins, tyrosine-rich proteins, lysozymes, and peptides (Ellison 1979; Barka 1980; Bradley 1991). During grooming, saliva functions much like soap and breaks down organic chemicals stuck to fur, feathers, and skin as well as the lipids that were applied earlier to keep surfaces soft and pliable but that have now outlived their usefulness. In doing so, saliva breaks apart large organic molecules into smaller ones, some of which are volatile enough to become odorants.

Other sources of odorants emanating from birds and mammals are the parasites, bacteria, yeast, and other microbes that live both in their digestive systems and on their outer surfaces. These microbes produce odors, especially when growing under anaerobic conditions that exist in areas where oxygen is absent or limited. Under such conditions, anaerobic organisms obtain energy by breaking down organic chemicals into their simpler parts, such as fatty acids, phenols, ammonia, and other volatile chemicals. Both birds and mammals possess many sites where anaerobic microbes can thrive; microbially derived odors include halitosis or bad breath, intestinal gases, and odors coming from infected wounds (Albone et al. 1977). Cook et al. (1971) noted that deer fawns that are ill and have diarrhea, elevated body temperatures, or discharge of bodily fluids (such as mucous or blood) produce a stronger odor than healthy fawns, and that ill fawns are more likely to be located and killed by coyotes (*Canis latrans*) than healthy ones.

Skin cells have a finite life, and dead ones are constantly replaced by new cells created from the lower layers of the skin. The result is that the outer skin surface is covered with dead and dying skin cells. The human body sheds about 40,000 skin cells each minute; they are shed as single cells and as small flat flakes containing numerous dead cells. Attached to them are staphylococci bacteria (Marples 1969; Syrotuck 1972) that feed on the dead cells. These cell flakes are so large and heavy that they fall out of the air quite rapidly compared to organic chemicals. The flakes lie on the ground much like fine dust until disturbed by turbulence, such as the breath of a sniffing dog. When this happens, the flakes become airborne, and some will be inhaled. The flakes themselves are too big to cross the mucous membrane and are therefore not odorants. However, bound to each flake are odorants associated with the skin or produced by the bacteria that degrade skin, and these bacteria produce their own unique odorants (Syrotuck 1972). Birds, of course, lose not only skin cells, but also small pieces of feathers, and there are feather-degrading bacteria, such as *Bacillus licheniformis*, that release their own unique metabolic odorants (Burtt and Ichida 2004).

When discussing the production of odorants by an animal, the most appropriate measure is the microgram (µg) of each type of odorant produced within a specified period of time. It is not useful to lump all odorants together into a single value because predators only use a few of them to identify prey, and their ability to detect odorants varies greatly among odorants. Before an odorant can be detected, it must be suspended in air; hence, when we discuss detectable levels of an odorant to a specific predator, the most appropriate measure is micrograms of odorant/liter of air.

Using odors to detect differences between species or individuals

All individual animals except for identical twins have a unique odor that is genetically determined. Animals have a keen ability to use odors to distinguish between individuals of their own species and even identify by odor whether they are genetically related to another individual (Tegt 2004). But, can they identify individuals of other species by smell alone? Domestic dogs can identify individual people based entirely on their odor (Settle et al. 1994). They can even use odor to distinguish between people who are genetically related to each other, but they cannot distinguish between identical twins (Kalmus 1955; Settle et al. 1994). Dogs store knowledge of a person's odor in their long-term memory so that they can still individually recognize a person who has been absent for over a year. Hunting dogs can use odor to distinguish between individual birds. Most hunting dogs, for instance, show little interest in the scent of a bird that they have already retrieved but great interest in the scent of a new individual (Figure 1.8).

Many mammals use their urine to communicate their presence to conspecifics. When females are in estrous and receptive to mating, odorants in the female's urine communicate this to males along with the identity of the female. Territorial animals urinate along the borders of their territories to mark them. Olfactory cues in the urine inform other animals both that the territory is occupied and which individual occupies it (Erlinge 1977). Some animals can even determine if they are genetically related to the territory holder by smelling its feces or urine (Tegt 2004). Of course, predators can also smell the feces and

Figure 1.8 Hunting dogs can individually recognize birds by their odor and show little interest in those that they have already retrieved.

urine of other species and determine which species produced them. For instance, trained dogs can use odor to distinguish between the feces produced by similar species, such as a kit fox (*Vulpes macrotis*) and red fox (*Vulpes vulpes*), or between a black bear (*Ursus americanus*) and a grizzly bear (*Ursus arctos*) (Smith et al. 2003). Such information can help a predator decide where to hunt for prey.

Likewise, prey can detect the presence of predators by their body odor and avoid those areas where the scent of a particular predator is located (Lindgren et al. 1997; Caro 2005). For example, southern flying squirrels (*Glaucomys volans*) stayed out of cavities that contained the scent of predators (Borgo et al. 2006). Populations of montane vole (*Microtus montanus*) declined in areas treated with anal gland compounds of the short-tailed weasel (*Mustela erminea*) (Sullivan et al. 1988). Pocket gophers (Geomyidae) avoided traps contaminated by secretions from the anal scent gland of mustelids (Sullivan and Crump 1986).

Prey also can identify the urine and feces of predators; they stop feeding and scent marking, and they vacate areas where predators urinate or defecate (Sullivan et al. 1988; Caro 2005). Such behavior may help prey avoid predators because predator urine attracts conspecifics (Whitten et al. 1980; Clapperton et al. 1988, 1989). These avoidance responses are elicited only by urine and feces of predators and not by urine or feces of herbivores. As an example, urine from a wolf, coyote, fox, bobcat (*Lynx rufus*), and lynx (*Lynx canadensis*) suppresses feeding behavior in snowshoe hares (*Lepus americanus*), but deer urine does not (Sullivan et al. 1985; Sullivan 1986; Sullivan and Crump 1986; Lindgren et al. 1997). The response of herbivores to predator odors appears to be innate. Muller-Schwarze (1972) showed that naïve, hand-reared mule deer (*Odocoileus hemionus*) avoided predator odors. Gorman (1984) discovered that voles on Orkney Island avoided traps containing anal gland secretions from stoats (*Mustela erminea*) even though stoats have been absent from the island for over 5000 years.

Can animals hide from olfactory predators by changing their odor?

If animals cannot stop releasing odorants, then can they disguise their odor so that they will not be recognized by a predator? It would be advantageous for an animal if its odor was unique and unlike the odor of conspecifics and other animals that an olfactory predator was hunting. Likewise, predators would benefit if prey could not recognize their odor or the odor of their feces and urine. The ability of either predators or prey to disguise their odor, however, is probably limited. Every animal emits hundreds of different odorants, and predators can sort through all of them to find the specific odorant or combination of odorants that is unique to that particular prey. Hence, an animal would not be able to identify which set of odorants a particular predator is using at a particular moment to detect it. Therefore, it would not know how to modify its odorants so that it could escape detection.

Can animals hide from olfactory predators by masking their odor with another, overpowering one?

Whether animals can hide from olfactory predators by masking their odor with another, overpowering one is an interesting question that has not been adequately tested. As most dog owners can attest, dogs like to roll in dead carcasses and get the odor all over them. I often wondered if this helps to mask their odor, which would be a virtue when stalking prey. However, foxes also like to roll in carcasses, and Budgett (1933) reported that after foxes had done so that they were much easier for hounds to track.

There is some evidence that the use of feces from other species may make it harder for olfactory predators to locate bird nests. Some burrowing owls (*Athene cunicularia*) carried livestock dung into their burrows and line their nests with it (Figure 1.9). Owls that do so are less likely to have their nests depredated by American badgers, a predator that used olfaction to locate these nests (Green and Anthony 1989).

Using one's own feces to mask one's scent, however, would be counterproductive if the feces betrays one's presence to olfactory predators. It is probably for this reason that many birds with altricial young carry away their chicks' feces from the nest. In fact, Petit et al. (1989) showed that nest predation rates increase when these feces remain too close to a nest. Still, the overpowering smell of fresh feces may distract an olfactory predator that is using the odor of a nest to locate it. In this regard, it is worth noting that Sandwich terns (*Thalasseus sandvicensis*) deposit large quantities of feces close to their nest (Caro 2005). Ducks also exhibit an interesting behavior when flushed from their nests; some, but not all, defecate as they fly away so that the defecation lands either on or within a meter of their nest. The reason for this behavior is unknown. One hypothesis is that this behavior makes the nest harder to find by masking the odor of the nest. The second hypothesis is that it makes the eggs less palatable, and the third is that it has no effect on predators but is merely a reflex action on the part of the fleeing duck.

Townsend (1966) tested the first hypothesis by examining the nesting success of 11 duck species on Egg Lake in Saskatchewan, Canada. Over half of the nests were lesser scaup (*Aythya affinis*) or ring-necked ducks *(Aythya collaris)*. When first flushed from the nests, 30% of 196 hens defecated. When these nests were rechecked 7 to 10 days later, 12% of the 58 defecated nests had been depredated, as had 20% of the 138 nondefecated nests. Although predation rates were lower at nests where hens defecated when flushed, this difference was not statistically significant because of a small sample of depredated nests. Hammond and Forward (1956) also found no difference in predation rates of artificial nests when half of the eggs were scented with duck feces and half left untreated. In a third study, Clark and Wobeser (1997) actually found that artificial nests scented with duck feces suffered higher predation rates than unscented artificial nests. Furthermore, efforts by wildlife biologists to reduce mammalian predation of artificial or natural nests by scenting the nests with a mustard-kerosene mixture, perfume, doe-in-rut scent, naphthalene mothballs, or pulegone have failed (Hammond and Forward 1956; Gawlik et al.

Figure 1.9 Burrowing owls line their nesting burrows with dung. (Courtesy of Ray Kirkland and the U.S. Fish and Wildlife Service's Bear River Migratory Bird Refuge.)

1988; Whelan et al. 1994; Conover and Lyons 2005). Hence, data do not support the hypothesis that ducks defecate to distract olfactory predators.

There is more evidence for the second hypothesis that defecating on eggs makes them less palatable. Swennen (1968) reported that incubating common eiders (*Somateria mollissima*) produce a horrible-smelling excrement that they squirt over their eggs when they are flushed from their nest. The same excrement is not produced by nonincubating ducks. The author gave some food soiled with the excrement to some captive ferrets and rats and found they would not eat it but would eat food soiled with excrement from nonbreeding eider.

Yet, when it comes to an overpowering odor, skunk musk quickly comes to mind because there are anecdotal reports that a strong whiff of skunk musk causes a temporary loss of smell (Figure 1.10). If true, olfactory predators should have a harder time locating nests in areas where skunks have sprayed musk. To test this, I located 57 artificial nests along the flood plain of the Bear River in Cache County, Utah, on land owned by Utah Power. Crested wheatgrass, reed canarygrass (*Phalaris arundinacea*), smooth brome (*Bromus* spp.), and cattails (*Typha* spp.) dominated the sites. The artificial nests were made by simply placing a brown, medium-size chicken egg on the ground. Because I hypothesized that the skunk musk would be effective only against olfactory predators, I wanted to make it easy for olfactory predators to find these nests, so I smeared the egg with egg white to give it a strong egg scent. I also covered the egg with dead grass collected from the nest's vicinity so that it would be less conspicuous to visual predators such as black-billed magpies (*Pica pica*). I then randomly selected half of the nests and placed 2 to 5 ml of

Figure 1.10 Hognose skunk (*Conepatus leuconotus*). (Courtesy of U.S. Department of Agriculture's Wildlife Services.)

skunk musk on the edge of each. The other half were left untreated as controls. I checked all nests daily for 4 days and noted whether the eggs had been depredated.

After 24 hours, few eggs from either treated or untreated nests were depredated. From then on, however, nests that had been treated with skunk musk suffered a much higher predation rate than untreated nests (Figure 1.11). Clearly, the odor of skunk musk did not overpower the egg odor and prevent olfactory predators from finding the nests. The puzzling question is, why did skunk musk cause an increase in nest predation? One hypothesis is that skunk musk might attract other skunks, but if so, then predation rates at treated and untreated nests should have been different starting on the first day. A more likely explanation is that local predators learned that a nest was located where there was a strong scent of skunk musk. Once they learned this association, they probably sought out and easily found the treated nests; the untreated nests were harder to detect.

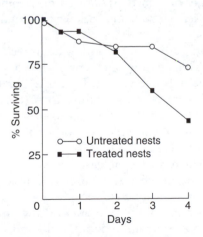

Figure 1.11 Percentage of artificial nests, containing a single chicken egg that survived over a 4-day period when half of the nests were treated by placing 2-ml skunk musk by them and the other half of the nests were untreated.

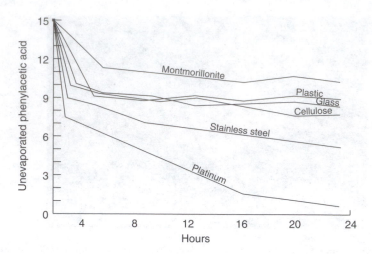

Figure 1.12 Impact of six different types of surfaces on the evaporation rates of phenylacetic acid at 0% relative humidity and 20°C. The *y* axis shows the amount of phenylacetic acid that has not yet evaporated. (Adapted from Regnier, F.E., and M. Goodwin, in D. Muller-Schwarze and M.M. Mozell, Eds., *Chemical Signals in Vertebrates*, Plenum Press, New York, 1977, pp. 115–133, and used with permission of Springer Science and Business Media.)

Factors influencing the evaporation rate of odorants

When a drop of odorant is placed on a solid object, most of the odorant evaporates within minutes or hours, but some of the odorant molecules become bound to the surface and may not evaporate for days, causing the scent to linger. The substrate on which an odorant is located has a great impact on the emission or evaporation rate of an odorant. A smooth surface (stainless steel or platinum) has few binding sites for odorants, and odorants on its surface evaporate quickly. In contrast, rough surfaces (for example, clay) have so many binding sites that most of the odorant may still be present 24 h after placement on the surface. Most natural surfaces on which an odorant may come into contact are rough (for example, feathers, fur, skin, leaves, soil, and rocks) and contain considerable binding sites (Figure 1.12).

Odorants rarely occur in pure form but rather are mixed together with other volatile and nonvolatile chemicals. This mixture of chemicals has a major impact on the evaporation rate of an odorant, usually by slowing it (Figure 1.13). Humidity also has an impact on the evaporation rate of polar odorants bound to surfaces because water molecules, which also are polar, compete with the odorants for the binding sites. In doing so, the water molecules speed up the rate at which the odorants are released from the binding sites and are free to evaporate (Figure 1.14).

When a drop of odorant has been released in an enclosed space under constant conditions, equilibrium will occur when the number of molecules evaporating equals the number leaving the gas phase. How much of the odorant dissolves before this equilibrium is reached depends on the nature of the odorant and the ambient temperature (Regnier and Goodwin 1977). The more volatile the odorant, the more it will evaporate before the equilibrium point is reached. Also, the higher the ambient temperature, the more the odorant will evaporate before equilibrium. This explains why things dry faster on warm days. When there is no breeze, the air next to a wet rag will become saturated with water molecules, and the evaporation rate of water on the rag will decrease as the equilibrium point is approached. However, when there is a breeze, air next to the wet rag is constantly

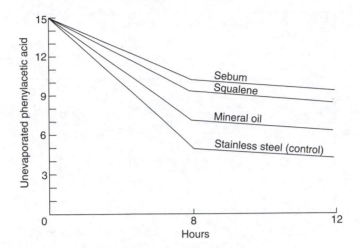

Figure 1.13 Impact of three different nonvolatile liquids (mineral oil, squalene, and sebum) on the evaporation rates of phenylacetic acid from a stainless steel surface at 0% relative humidity and 20°C. (Adapted from Regnier, F.E., and M. Goodwin, in D. Muller-Schwarze and M.M. Mozell, Eds., *Chemical Signals in Vertebrates*, Plenum Press, New York, 1977, pp. 115–133, and used with permission of Springer Science and Business Media.)

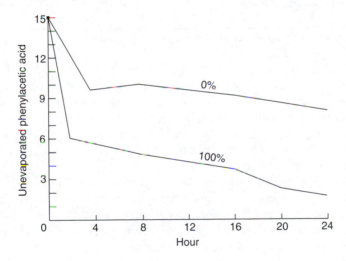

Figure 1.14 Effect of relative humidity (0 and 100%) on the evaporation rates of phenylacetic acid from a cellulose surface at 20°C. (Adapted from Regnier, F.E., and M. Goodwin, in D. Muller-Schwarze and M.M. Mozell, Eds., *Chemical Signals in Vertebrates*, Plenum Press, New York, 1977, pp. 115–133, and used with permission of Springer Science and Business Media.)

replaced by drier air, so equilibrium is never reached. Hence, the wet rag will dry quickly when there is a breeze. The same principle applies to the evaporation of all odorants.

Thus far, we have discussed the processes that increase the evaporation rate of volatile chemicals. However, even large organic chemicals can become suspended in the air when they obtain enough kinetic energy from an outside source, such as wind. The splash from a falling raindrop also hurls large organic molecules that had been lying on a surface into the atmosphere (Figure 1.15) (Gregory 1973). Large organic molecules also become airborne whenever an animal scratches itself, brushes against something, coughs, sneezes, or makes a loud noise. The convective heat given off from the surfaces of warm-blooded

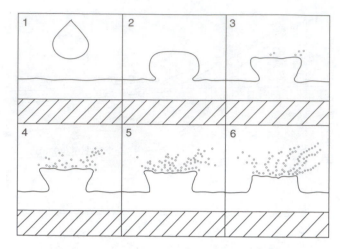

Figure 1.15 A raindrop falls with sufficient momentum that its splash can carry large organic chemicals, cells, and viruses into the atmosphere.

animals can generate air currents adjacent to the skin that have sufficient energy to launch heavy organic chemicals into the air. These large molecules may only remain suspended in the air for short periods of time (seconds or minutes); small volatile molecules may remain suspended for days.

Sidebar 1.1: Hiding Kills from Scavengers

One difficulty for predators is protecting their kill from scavengers. The best way to do this for the biggest predator in the area is to stay by the carcass and scare other scavengers away. This method is commonly employed by wolf packs (Mech 1970). Guarding a carcass, however, is not an option for smaller predators that can be driven from the prey by bigger predators or for a predator foraging alone. In such cases, the predator often tries to hide the food from scavengers. Wolves, coyotes, and red foxes cache food by digging holes, placing the food inside it, and then covering the hole with dirt. Wolves then use their noses to pack the dirt tightly in the hole (Mech 1970). This behavior of tightly packing the dirt would not help hide the carcass from visual predators but would aid in hiding the food from olfactory predators because there are few air pockets in the soil that would allow the release of odorants into the atmosphere.

Losing kills to scavengers is a particular problem for predators in the Serengeti and other parts of Africa where there are many predators in terms of both densities and species diversity (Figure 1.16). One method used by Serengeti predators to hide a carcass from both visual and olfactory predators is to bury it, but this is impractical for large carcasses, such as a dead wildebeest (*Connochaetes taurinus*). For large carcasses, some hyenas remove pieces of meat and carry them far from the carcass to bury them across a wide area. Other hyenas have developed the ingenious method of taking large pieces of meat and dropping them into waterholes, ponds, or other water bodies (Kruuk 1972; Mills 1990). This method is effective against olfactory predators because submerged meat will release few odorants, which would first have to dissolve into the water and then evaporate into the air.

Figure 1.16 Hyenas must be able to hide carcasses from both visual and olfactory predators. (Courtesy of Johan du Toit.)

Movement of odorants through the atmosphere

Three forces control the movement of airborne odorants: diffusion, convection, and gravity. Diffusion results from Brownian motion or the random collision of molecules. These random collisions have the effect of moving odorants from areas of high concentration to those with low concentration. Diffusion moves odorants through the air at velocities of less than 1 m/hour. Yet, if a bottle of perfume is opened in an enclosed room, its odor permeates the entire room within minutes. This movement, however, results not from diffusion but from convection, the bulk movement of air. Even in a completely enclosed room, convective forces are more important than diffusion in moving odorants through the air. This is certainly true in the natural environment.

People consider air to be still when its velocity is less than 4 to 6 km/hour because below this velocity they cannot feel wind blowing against their skin, and the wind does not have enough force to move vegetation. However, even when the air seems still to humans, the force of convection moves odorants through the air at speeds thousands of times faster than the forces of diffusion or gravity. Hence, the forces of diffusion can be ignored when considering the movements of odorants through the air because they play an insignificant part compared to the forces of convection.

Convective forces result from differences in air temperature and pressure, which in turn result because not all surfaces on the earth are evenly heated by the sun. Convective forces can occur on a minute scale, such as when an air current a centimeter in length occurs between one part of a leaf in direct sunlight and another part that is shaded. They can also occur on a grand scale and result in the movement of air over thousands of miles, producing weather phenomena such as cold fronts, hurricanes, high-pressure centers, and jet streams.

Given enough time aloft, particles will settle out of the air because of gravity, with the largest and densest particles sinking fastest (Table 1.2). The same is true for molecules; those with the greatest molecular weight will, in general, be the first to settle out of the air. This downward movement of odorants because of gravity is imperceptibly slow

Table 1.2 Impact of Particle Sizes on Their Terminal Velocity (U) Caused by Gravity and the Rate (X) at which Brownian Motion Causes a Cloud of Them to Expand in Size

Particle diameter (µm)	U (cm/second)	X (cm/second)
20	1.3	1.5×10^{-4}
10	3.3×10^{-1}	2.2×10^{-4}
2	1.3×10^{-2}	5.0×10^{-4}
1	3.3×10^{-3}	7.1×10^{-4}
0.2	1.3×10^{-4}	1.6×10^{-3}
0.1	3.3×10^{-5}	2.2×10^{-3}

Source: Data from Dimmick, R.L. and A.B. Akers, *An Introduction to Experimental Aerobiology*, John Wiley and Sons, New York, 1969.

compared to the force of convection. Nevertheless, the force of gravity on odorants plays an important role in helping olfactory predators locate prey. Because heavier odorants will sink faster than lighter ones, predators can compare the relative concentration of odorants in an odor plume to determine how long the odorants have been aloft and therefore how far away they are from the source of the odorants.

The olfactory concealment theory

Every time an animal walks, it increases its risk of falling victim to an olfactory predator because by moving it is laying down a depositional odor trail that informs olfactory predators not only that it passed by a particular place, but also how long ago it was there and in which direction it was heading. Olfactory predators can then follow the depositional odor trail to stalk the animal and kill it. The next chapter deals with how these depositional odor trails are created and how olfactory predators use them.

Stationary animals are constantly emitting a stream of airborne odorants that are carried downwind. Such airborne odorants pose a much greater danger to animals than depositional odor trails because animals remain motionless when they sense danger, and they cannot stop emitting airborne odorants even when standing still. Hence, airborne odorants are more useful to olfactory predators in detecting and locating prey. In this book, I develop an olfactory concealment theory about how olfactory predators use depositional odor trails and airborne odorants first to detect the presence of prey and then to determine its location.

The olfactory concealment theory states that the ability of predators to detect and locate prey using olfaction varies with time, location, and atmospheric conditions. This variance is caused by updrafts, which cause the airborne odorants to rise higher than a predator's nose, and atmospheric turbulence, which makes airborne odor plumes meander and odorant concentrations vary unpredictably within the plume. Turbulence also causes airborne odorants to disperse more rapidly and shortens the distance over which an olfactory predator can detect prey. The times and places where updrafts and turbulence occur are consistent and predictable.

The olfactory concealment theory further states that animals that are vulnerable to olfactory predators should hide from these predators by positioning themselves in areas where updrafts and turbulence occur, and those animals that do so will have a higher probability of surviving than those that hide elsewhere. According to the olfactory concealment theory, animals should also engage in dangerous activities, such as foraging, during those periods when atmospheric conditions favor updrafts and turbulence.

The olfactory concealment theory offers an explanation for various predator and prey behaviors and experimental results that heretofore have not made sense. It also provides a framework that can guide future research. Before we can begin to test this theory, however, it is necessary to learn how olfactory predators use depositional odors to track prey (Chapter 2) and how they use airborne odor plumes to detect the presence of prey and then locate it (Chapters 3 and 4).

chapter two

Detecting and locating prey through depositional odor trails

There are two ways predators can use odorants to locate prey. The first is airborne odors, and the second is depositional odors that lie on the ground or other surfaces. The latter are chemicals sloughed off by an animal as it moves along the ground and airborne chemicals that have settled out of the air column and lie on surfaces. If undisturbed by airflow or precipitation, then deposited chemicals may lie undisturbed on a surface for hours or days, allowing an animal to detect the earlier passage of another. However, predators cannot smell depositional chemicals as long as these chemicals remain on a surface; instead, the chemicals must be resuspended into the air to be detected. A predator can accelerate this resuspension of odorants by placing its nose close to the ground and inhaling rapidly. This creates an air current along the surface that causes the odorants to be resuspended and inhaled. Beagles can use depositional odor trails to track and run rabbits for miles (DiBenedetto 2005), and trailing dogs can follow the path made hours earlier by a person (Johnson 1977; Lowe 1981). Such formidable olfactory abilities are used by predators to locate and track their prey.

Creation of depositional odor trails

As an animal moves through an area, it leaves a complex odor trail. As an animal's feet strike the ground and as its body brushes against vegetation and other objects, it leaves a trail of odorants to which is added the odor of crushed vegetation caused by the animal's passage and the odor of disturbed soil. This is called the *path trail* (Syrotuck 1972), *track scent* (Johnson 1977), or *contact odor trail* (Hepper and Wells 2005). As the animal passes, flakes of skin or feather fragments are also sloughed off its body along with organic molecules. The larger, heavier skin flakes, consisting of thousands of cells, will be carried a little downwind before they settle on the ground, and the heavier flakes will often overlay the path trail. Overlying this path of skin flakes and extending even farther downwind from it will be a trail of organic molecules of high molecular weight and low volatility that were emitted by the animal as it passed by and quickly settled out of the air and onto the ground. Extending farther downwind will be other odorants of lighter molecular weight (Figure 2.1). The skin flakes and odorants lying on the ground down-wind of the path trail combine together to make up the body odor trail (Johnson 1977). The body odor trail and the path trail both make up the depositional odor trail.

Syrotuck (1972), Johnson (1977), and Hepper and Wells (2005) observed that there was a difference in how dogs follow an animal's depositional odor trail. Tracking dogs cued in on the animal's footprints (that is, the path trail) and followed them exactly. These dogs kept

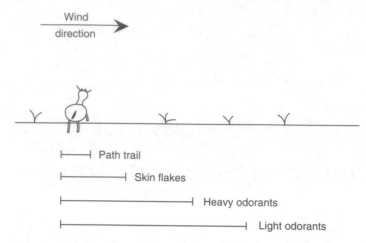

Figure 2.1 The different types of depositional odor trails left by a passing animal.

Figure 2.2 This hunting dog is following the odorant trail of a ring-necked pheasant and is moving a couple of meters downwind of the actual path taken by the bird.

their noses within 5 cm of the ground, followed the same path made by the animal, and did not deviate from it. In contrast, trailing dogs followed a trail with their heads 10 to 20 cm above the ground, and they did not follow the animal's tracks exactly but tended to travel downwind of the path trail (Figure 2.2) (Hepper and Wells 2005). They followed the body odor trail rather than the path trail and were prone to taking shortcuts across corners or to overshooting the path trail when it turned sharply (Syrotuck 1972; Johnson 1977).

Determining how long ago a trail was created

The composition of both the path trail and the body odor trail changes over time because some odorants decompose or disperse more rapidly than others. The path trail is usually detectable for much longer than the body odor trail (Syrotuck 1972; Johnson 1977;). Dogs and other mammalian predators can use these changes to determine the age of a trail. Such information is crucial in allowing a predator to determine when a trail is worth following (that is, how far ahead is the quarry). For instance, foxhounds (Budgett 1933), wolves (Mech 1966), and bird dogs (Syrotuck 1972; personal observation) will not follow an odor trail unless it is fresh.

Predators can also assess when they are getting close enough to their quarry that they need to be prepared for it to flush. Dogs bred to point (for example, German short-haired pointers) can tell when they are getting close to a bird by smell alone and freeze into a pointing position. Based on changes in its scent, experienced beagles can detect not only when they are getting close to their quarry (usually a rabbit), but also when it is tiring (DiBenedetto 2005).

Sidebar 2.1: Lessons from 1000 Years Ago about How to Throw Off Dogs from Your Depositional Odor Trail

According to the Icelandic narrator Snorre Sturlason, who lived in the 13th century, the Swedes captured and imprisoned two Norwegians in a deep hole in a pigpen during 1026. The Norwegians knew that if they escaped, the Swedes would find their trail and track them down using dogs. To outwit the dogs, the crafty Norwegians tied reindeer hooves pointing backward to the bottom of their boots. Sure enough, the prisoners' trail was soon located. However, when the Swedes released the dogs, the dogs were fooled because the reindeer hoof prints pointed the wrong way, and the dogs led the Swedes back to the pigpen (Steen and Wilsson 1990).

Determining the direction of an odor trail

When a predator comes across an odor trail, it has to be able to tell which way the animal was traveling so the predator will not follow the trail in the wrong direction; exactly how the predator does so has long puzzled scientists. According to legend (Sidebar 2.1), dogs detect the direction an animal is moving by the way its foot is pointing. Another similar possibility is that when people walk, their heel hits the ground before their toes, and dogs may be using this to determine the direction of travel. While Snorre Sturlason's story (Sidebar 2.1) has endured through the ages, dogs do not determine which direction a person is moving by noting the direction a footprint is pointing or determining which end of a footprint is the heel or toe by the slight time difference when they strike the ground. When people walk backward so that their toe hits the ground before their heel, dogs are not fooled and correctly follow the person's direction. Instead, dogs detect the direction an animal or person was traveling by detecting differences in odorant concentrations among consecutive footprints that resulted because odorants in the older footprint have had a longer time to disperse (Steen and Wilsson 1990; Hepper and Wells 2005). In a clever experiment using carpet squares that each contained a single footprint, Hepper and Wells (2005) showed that dogs can tell which way a person was walking using the olfactory information in the path trail. They also showed that dogs need five footprints to determine the direction the person was moving. When given three footprints, the dogs guessed the wrong direction half the time (Figure 2.3).

Detecting the direction someone was walking requires a keen ability to detect small differences in odorant concentrations between consecutive footprints. For instance, in the Canadian Kennel Club's test for the title Tracking Dog Excellent, the dog must be able to follow a trail that is at least 3 hours old (Johnson 1977). The difference in odorant concentration between five consecutive footprints made by a person walking 3 hours earlier is about 2×10^{-4}, so dogs awarded the title Tracking Dog Excellent can certainly detect these slight differences.

It also turns out that the space between footprints is important for a dog to determine which way a person was moving. When a person drags his or her feet on the ground,

Figure 2.3 Dogs can tell which direction a person was traveling by smelling five footprints. (Courtesy of John Shivik.)

tracking dogs often make mistakes and head the wrong way (Steen and Wilsson 1986). For the same reason, dogs cannot determine the direction a bicycle was traveling (Steen and Wilsson 1990).

Sidebar 2.2: Lessons from Hollywood about How to Keep Dogs from Following a Depositional Odor Trail

Most people today know little about how to thwart an olfactory predator from tracking them. They probably learned what they do know from watching movies about prisoners eluding the prison's bloodhounds or slaves escaping from the dogs of bounty hunters. What is repeated in movie after movie is that there are three methods to elude dogs from following a person's trail. The first is to walk upstream or downstream in the middle of a stream or float down a river. The second is to walk over bare rocks. The third is to backtrack; this involves stopping at some point and walking backward, ideally placing your feet in the exact same place where they first struck the ground. The really smart Hollywood heroes like Walt Disney's Davy Crockett or Daniel Boone would then jump up and grab a tree branch and climb along it for a while so that their feet did not touch the ground. According to Hollywood, these techniques were all effective at fooling any dog following a trail (at least for the good guy).

Impact of environmental conditions on depositional odor trails

Several environmental factors influence the strength of a depositional odor trail and the ability of predators to follow it. One factor is the surface over which the quarry (the person or animal that is making the trail) is moving. The depositional odor trail will both be stronger and last longer on surfaces that are vegetated than on those over bare ground. Trails over paved roads are particularly ephemeral. Budgett (1933) reports that hounds sometimes lose track of foxes when they cross newly plowed fields and clay soil sticks to their paws. Although the clay may slow them down, the clay shoes prevent their paws from making contact with the ground and reduce the amount of odorants they leave behind. In contrast, wounded animals are easily tracked by the smell of their blood.

Another environmental factor that has an impact on depositional odor trails is wind velocity at the time the depositional odor trail was produced. In still air, the path trail, skin flake trail, and body odor trail will be narrow and overlap, and odorants will be concentrated. Under such conditions, depositional odor trails will be easier for a predator to follow and will persist longer. Windy conditions when predators are trying to follow an odor trail will also make it difficult to follow the trail (Budgett 1933; Johnson 1977; Lowe 1981). However, it is easier for dogs when they are following a trail into the wind rather than downwind because in the former situation, odorants from the trail in front of them are blown to them.

Another important factor is whether ground and vegetation surfaces were dry or wet when the depositional odor trail was created. Tracking dogs work better if the trail is laid down when surfaces are moist or if there is light drizzle than when surfaces are dry (Budgett 1933; Syrotuck 1972; Johnson 1977; Lowe 1981). Spotted hyenas can also easily follow an odor trail when the ground is damp (Mills 1990). However, a hard rain either when a trail is laid down or afterward may obliterate it (Budgett 1933; Clifford 1964; Johnson 1977; Lowe 1981). A light rain after a trail is laid, however, seems to freshen the trail and makes it easier to follow (Budgett 1933; Johnson 1977). Dogs bred to hunt birds (pointers and flushers) work best when the ground is damp but have difficulty finding birds in a pouring rain.

Environmental conditions also influence how long an odor track will persist before the scent becomes too weak for a predator to follow. As surfaces become wet from rain, dew, or high humidity, water displaces many odorants from surface-binding sites, allowing them to evaporate and enter the air column. This accounts for the strong aroma of a wet meadow or a forest after a summer rain. Such conditions are ideal for dogs to follow trails. For instance, search-and-rescue dogs can find people faster when the relative humidity is high (Shivik 2002). English foxhounds have an easy time following trails when it is foggy or when conditions are damp (Budgett 1933). Bird dogs and tracking dogs have an easier time finding game or staying on an odor trail under such conditions as compared to hot, dry, or sunny conditions. Gutzwiller (1990) surmised that this was because dry, sunny weather is not conducive to bacterial degradation of skin flakes, and that the odorants released by these bacteria were the primary ones used by dogs to locate birds. It seems more likely that such conditions cause odorants to decompose or evaporate quickly so that odor trails do not last long.

However, moisture also hastens the destruction of a depositional odor trail because more of the odorants lying on the surface become resuspended in the air. This will make it harder for a predator to follow a depositional odor trail once these conditions have passed because there will no longer be a large concentration of odorant molecules on the surface. For instance, trailing dogs perform better when a light drizzle has freshened a trail that was created earlier, but their performance drops once the ground has dried again (Johnson 1977).

After a depositional odor trail has been produced, wind and high temperatures will cause it to fade by increasing the rate at which odorants evaporate. In addition, high temperatures and sunlight will cause many odorants to decompose (Lowe 1981) and may kill the bacteria on skin flakes (Gutzwiller 1990). Under such conditions, a depositional odor trail can disappear in as few as 15 minutes (Doving 1990). In contrast, depositional odor trails can persist for up to 48 hours on overcast days or in thick vegetation. This may be why many bird hunters report that their dogs have more difficulty finding birds when temperatures rise above 25 to 30°C than when it is cooler (personal observation). Foxhounds (Budgett 1933) and bloodhounds (Lowe 1981) have more difficulty trailing their quarry on days when it is hot, sunny, or windy.

Tracking dogs (Johnson 1977), foxhounds (Budgett 1933), and bloodhounds (Lowe 1981) perform best when the ground is warmer than the air, presumably because the trail scent rises. Budgett (1933) noted that because air changes temperatures faster than the ground, the latter is warmer than the air at sunset, at night, and when there is a sudden fall in air temperature. These are the conditions when predators will have the easiest time following a trail. For the same reason, predators have a harder time following a track when the air temperature is rising because the ground temperature lags behind it and hence will be cooler.

If snow falls after a trail is laid down, then the effect of the snow cover on trail odorants varies. If snow falls before the ground is frozen, then the snow helps insulate the ground, keeping it warm, so that hounds, wolves, and other predators have an easy time following a trail (Figure 2.4). However, if snow falls on frozen ground, few odorants escape the snow cover, and hounds will not be able to follow the trail (Budgett 1933).

These results suggest that the best time for predators to try to locate prey using depositional odor trails would be on days when it is cool and humid (Budgett 1933; McCartney 1968). The best time to hunt would be at sunset, at night, or around sunrise when the air is calm and the ground is likely to be moist and warmer than the air (Budgett 1933; Johnson 1977). The worst time for predators to track prey using depositional odor trails and therefore the best time for prey to be moving about would be during a hard rain or in the middle of a bright sunny day.

Figure 2.4 Tracking conditions are good when a light snow falls on unfrozen ground, and such conditions allowed these wolves to track their quarry. (Courtesy of U.S. Department of Agriculture's Wildlife Services.)

Figure 2.5 Wolves often hunt in large packs on Isle Royale National Park. (Courtesy of U.S. Department of Agriculture's Wildlife Services.)

Sidebar 2.3: Description of Wolves Locating Moose by Following Their Depositional Odor Trail

From aircraft, Dave Mech was able to observe wolf hunts on Isle Royale located in Lake Michigan where the wolves were using olfaction to locate their prey (Figure 2.5). During nine hunts, wolves located their prey by following depositional odor trails. The following is his description of one such hunt that took place on February 11, 1960 (Mech 1970, pp. 175–176):

The 16 wolves left a swamp and struck out into an open burn; they appeared to be on a fresh moose track. When 250 yards crosswind of three adult moose (two lying, one standing), they stopped and scented the air (5:15 p.m.). The first [wolves] lay on a ridge 200 yards from the moose for a minute, while the rest caught up. They continued along the trail, noses to the ground. Two wolves remained downwind and about 25 feet ahead of the trackers. All three moose were lying down, but when the first two tracking wolves got within 25 feet, they arose. Meanwhile the rest of the wolves caught up.

How good are predators at following a depositional odor trail?

Under ideal conditions, a tracking dog can follow a trail made days earlier (Johnson 1977). There are differences among species in how long their trails last. On sunny days, odorants from the trail of a fox last 20 to 30 minutes, hare 45 minutes, ring-necked pheasant (*Phasianus colchicus*) 3 hours, human 5 hours, red deer (*Cervus elaphus*) 12 days, and otter for more than 2 days (Budgett 1933). However, hounds can follow a person's trail 5 days after it was produced because plants crushed by the person's footsteps continue to give off odorants long after odorants left by the person have dissipated (Budgett 1933).

How far a predator can follow a depositional odor trail or how much time will pass before the odorants in the trail become undetectable is hard to answer for free-ranging predators because most are only interested in fresh trails rather than old ones. However, some data are available. Mills (1990) observed that spotted hyenas, a species that hunts mainly at night, followed depositional odor trails a mean of 0.9 km ($n = 11$), with the maximum distance 1.8 km; brown hyenas followed an odorant trail for a maximum of 1.4 km. Mills (1990) also noted that native trappers, when trying to catch a hyena, start a few kilometers from the trap's location and drag a carcass to the trap as a means of getting hyenas to come to the trap.

Behavioral tactics used by deer and hares to escape from tracking dogs

When running, red deer stags give off great quantities of scent that dogs can detect several hundred meters downwind (Geist 2002). When pursued by hounds, red deer commonly use three strategies to throw off their pursuers: running into a group of other deer, doubling back on their own tracks, and running into water (Geist 2002). Moose and caribou (*Rangifer tarandus*) also commonly move into water when trailed by predators (Figure 2.6).

White-tailed deer (*Odocoileus virginianus*) showed five distinctive methods to escape from hounds tracking them (Sweeney et al. 1971). These included (1) bedding in thick cover or deep snow (Figure 2.7); (2) running in a relatively straight course for a long distance; (3) running in a circuitous, zigzag pattern in which the deer often crossed its own trail; (4) joining a group of deer and then separating from the group; and (5) running through water.

Sweeney et al. (1971) observed hounds trailing white-tailed deer equipped with radio collars so that the scientists were able to ascertain the deer's behavior and the outcome of the pursuit on 65 occasions. Deer attempted to avoid detection by hounds 7 times by remaining bedded in thick cover, and on 4 of these occasions, the dogs failed to pick up the scent. Deer ran in a straight course to escape pursuing hounds 7 times, and the dogs lost the trail in all cases. There were 19 deer that ran in circuitous patterns, and 10 times the dogs lost the trail (Figure 2.8). Seventeen deer joined and later separated from a group of deer, and 14 times the dogs followed the group of deer rather than the radio-tagged deer. Deer ran through or swam across water 40 times, and this tactic was successful 38 times (Sweeney et al. 1971).

Figure 2.6 Moose often run or swim across water bodies when pursued by wolves. (Courtesy of U.S. Department of Agriculture's Wildlife Services.)

Figure 2.7 When deer are bedded in dense vegetation or deep snow, they will often remain motionless when pursued by hounds. (Courtesy of Tony DeNicola.)

Figure 2.8 Deer will often run in circuitous patterns and cross their own trail when pursued by hounds. (Courtesy of Tony DeNicola.)

When trailed by foxes or dogs, a hare will run forward for considerable distances. Often, it will turn back on the same track and then make a great leap at right angles to its former course (Budgett 1933). The hare then lies quietly in its spot of concealment. When the snow is deep, it may burrow into it so that only a small air hole betrays its location (Budgett 1933). Doing so reduces the amount of odorants released from the hare's fur, and the trailing predators often pass by without detecting it.

Locating home ranges using olfactory cues

Most animals have specific home ranges where they are located most of the time. It certainly would be in the best interest of a predator to locate the specific home range of an animal and concentrate its hunting there rather than searching for prey randomly. Predators can use olfactory cues to locate these home ranges. Certainly the presence of numerous depositional odor trails would be a clear sign to a predator of an animal's home range. For species that use pheromones to mark their territories, these odors also help predators identify occupied territories. In addition, the odors of an animal's feces and urine provide good evidence of an animal's home range.

Likewise, prey can identify the home range of a predator by noting the location of its feces and urine. Such information would be useful to prey animals by helping them avoid areas where the risk of running into a predator is high. Several studies have shown that many herbivores find the urine of predators that prey on them to be aversive (Conover 2001).

Of course, animals cannot stop urinating or defecating, so what can they do to reduce the information from their feces and urine that is available to olfactory predators? For one thing, they can always defecate or urinate in the same location (that is, they can create a latrine). Many species, such as horses (*Equus caballus*), do this. Some species, such as raccoons (*Procyon lotor*) and beavers (*Castor canadensis*), urinate in water (Figure 2.9). Other species bury their feces or urine in dirt or snow to reduce the amount of odorants that they release in the air.

Figure 2.9 Raccoons frequently urinate in water. (Courtesy of U.S. Department of Agriculture's Wildlife Services.)

> **Sidebar 2.4: An Instinctual Response of Humans to Elude Olfactory Predators**
>
> Few people today spend much time worrying that they are going to be stalked and killed by an olfactory predator. Yet, from an evolutionary perspective, it has not been that long ago when falling victim to a predator was a real threat to humans. Hence, it may not be surprising that humans still have instinctual behaviors designed to help us elude olfactory predators. Humans have an urge to urinate when they put their hands under running water. The most likely reason for this strange sensation is that humans with this instinctual response thousands of years ago were more likely to urinate in the water and therefore avoid leaving telltale olfactory signs of their presence in the area. Humans without this instinct were more likely to urinate on the ground and more likely to be detected and killed by olfactory predators. Hence, over many generations, humans with this instinctual response were more likely to survive and reproduce until almost all humans exhibit the same instinctual response to running water.

What prey can do to minimize their risk from depositional odor trails

Depositional odor trails pose more of a risk to a mammal than a bird because the latter can always take flight and bring its trail to an end. A mammal's depositional-odor trail ends only with itself. This, of course, is not good news for a mammal when a predator is following its trail. To reduce the risk of their depositional odor trail allowing them to be tracked, animals should turn often when foraging and search a small area intensely rather than forage by walking in a straight line. The reason for this is because the path of an olfactory predator is more likely to intersect a 1-km depositional odor trail that runs in a straight line than a 1-km depositional odor trail that constantly turns so that the ending point for the trail is close to its beginning. Animals can also minimize the risk of having a predator detect their depositional odor trail by moving during sunny days rather than at night because depositional odor trails do not last long on hot, sunny days.

When a predator is following an animal's depositional odor trail, the animal has several options to get the predator off its trail. First, the quarry can travel a long distance quickly so that some part of its trail will have time to disappear before the predator reaches it. This tactic is more likely to be successful on sunny, hot days when depositional odor trails do not last long. Second, the quarry can join a group of conspecifics and then separate from them or take a route so that its path crosses those of other conspecifics (Figure 2.10). Often, the predator will decide to follow another animal. Finally, the quarry can cross a water body. This tactic works best when the water body is wide and lacking in emergent vegetation. Otherwise, an olfactory predator can determine the direction taken by the quarry by the presence of its body odorants that have adhered to the vegetation of the banks of the water body.

What olfactory predators can do to maximize the usefulness of depositional odor trails

To increase their hunting success, olfactory predators should concentrate their olfactory abilities on depositional odorants that are species specific and are not under the voluntary control of the prey. Doing so will allow the predator to identify the animal that it is

Figure 2.10 When tracked by a dog, deer often mix with groups of other deer and then separate. This tactic was often successful because the dog would start to follow the wrong deer. (Courtesy of Tony DeNicola.)

following and will not allow the prey to stop producing the key odorant once it realizes a predator is tracking it.

Olfactory predators that hunt by following depositional odor trails should do so when environmental conditions favor the creation and retention of depositional odor trails. Such conditions occur when the ground is warmer than the air, the air is still, or the ground is moist from dew or a light shower. These events are most likely to occur at sunset, night, or sunrise. Conditions would be unfavorable on sunny days when sunlight is reaching the ground or during a hard rain.

An olfactory predator that has located a depositional odor trail must be able to answer three major questions if it is going to be successful in tracking and capturing the animal that made the trail. First, it must decide whether the trail is so old that it is not worth following. Second, it must decide in which direction the prey was traveling so it will know which way to head. Third, as it follows the trail, the predator has to be able to know when it is getting so close to the prey that it must get ready to pounce on the prey. All of these questions involve determining the age of the trail, and an olfactory predator can answer the questions by comparing odorant concentrations along the trail and noting which odorants make up the depositional odor trail and which ones have already disappeared. To do so, it must be able to detect a multitude of depositional odorants that vary widely in their volatility or half-life.

chapter three

Using airborne odorants to detect the presence of prey

All animals, even stationary ones, constantly release a stream of odorants into the air where these odorants are then carried downwind. Olfactory predators can use airborne odorant trails to detect their prey. Their ability to do so, however, varies with time and location. The sources of variation examined in this chapter are environmental variables that control the concentration of odorant molecules in the air. These environmental variables limit the ability of olfactory predators to detect the presence of prey. However, this is only the first part of an olfactory predator's task of foraging. It must also use these olfactory cues to determine where the prey or food source is located. How it does so is the subject of Chapter 4.

The challenge of using airborne odorants to detect the presence of prey

By emitting odorants into the air that are constantly carried downwind, even a stationary animal creates an airborne odorant trail (henceforth called an *odor plume*) that predators can use to detect its presence. To do so, the concentration of at least one of the prey's unique odorants, or more likely a unique combination of odorants, must be at a high enough concentration in the air at the predator's location for the predator to detect the odorant with its olfactory system. If odorant concentrations are below this critical threshold, then the prey is undetectable to the predator. This critical threshold concentration varies based on the odorant and the predator. But, as noted in Chapters 1 and 2, olfactory predators have developed remarkable abilities to detect biologically important odorants and can detect some of them at concentrations of parts per million. However, given enough distance between the odorant source and the predator, every odorant will drop to below detectable levels.

Animals seeking to avoid detection by olfactory predators want the odorants they are emitting into the atmosphere to drop to undetectable concentrations as soon as possible or, more precisely, in as short a distance as possible. To visualize why this is important, consider a bird emitting a stream of odorants that is floating downwind of it. If its odorants reach levels undetectable to a coyote within 200 m, it is at least twice as likely that a coyote's path will intersect the bird's odorant trail than if its odorants drift downwind only 100 m before they reach undetectable levels (Figure 3.1).

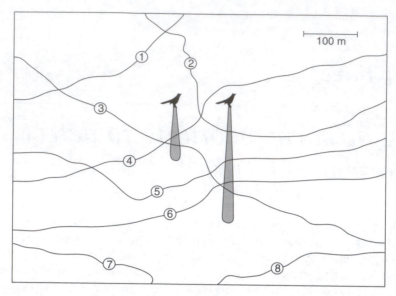

Figure 3.1 Odor plumes produced by two birds. One plume extends 100 m before it reaches unde-tectable levels, and another extends 200 m before doing so. The figure also shows the random paths taken by eight foraging olfactory predators. Two of these predators (Predators 3 and 4) detected the bird with a 100-m odor plume; four predators (Predators 2, 3, 5, and 6) detected the bird with the 200-m odor plume.

Impact of a steady wind on a predator's ability to detect an odor plume

Big game hunters are well aware that if they want to get close enough to their quarry to get a shot, they have to be downwind of it and approach so that the animal cannot smell them. Like big game animals, predators can detect only the odor plumes of prey that are upwind of them. Such is not surprising because odorants cannot move upwind.

It is sometimes frustrating for bird hunters when their dogs cannot find a dead bird that is in plain sight, but their dogs cannot find it because they are either crosswind or upwind of it. Free-ranging predators can also be oblivious to a large, visually conspicuous animal that is downwind of them. Dave Mech, the noted wildlife biologist, wrote, "I once watched a moose feeding undetected for 20 minutes while [it was] 100 yards downwind of a pack of 15 wolves" (Mech 1966, p. 119). Mills (1990) once observed a brown hyena passing 250 m upwind of a springbok (*Antidorcas marsupialis*) carcass without noticing it, although a similar carcass will attract a hyena from over a kilometer when the hyena is downwind of it. Skunks (Nams 1997) and raccoons (personal observation) often cannot detect prey when they are upwind of it.

How far can predators detect prey by sensing the quarry through its odor plume?

To answer the question of how far predators can detect prey by sensing the quarry through its odor plume, we need to distinguish between predators that are locating prey by following its depositional odor trail versus its odor plume. Predators are likely following depositional odor trails if they are following the same trail taken earlier by the prey. This is easier to determine when there is snow on the ground, and the prey's tracks are obvious. Mech (1970) used this technique to determine that wolves on Isle Royale National Park in Michigan detected moose by following their depositional odor trail 18% of the time and by detecting their odor plume 82% of the time ($n = 51$ observed hunts). Wolves were usually within 300 m of the moose when they detected its odor plume, but on one occasion they detected a cow and twin calves from over 2 km (Mech 1970).

Another way to identify that animals are following odor plumes rather than depositional odor trails is to examine the distances over which a carcass first attracted scavengers. Count Coutelux de Caneteleu, who observed the behavior of hand-reared wolves, noted that, under ideal weather conditions, wolves could detect a carcass when 5 km downwind of it (Budgett 1933). Mills (1990) determined that the mean distance from which spotted hyenas using just olfaction could detect six untouched carcasses was 3.2 km, and the maximum was 4.2 km. In all of these cases, the hyenas were downwind of the carcass when they detected it and immediately changed direction when they entered the odor plume of the carcasses. In contrast, when hyenas were upwind of a carcass, even a rotting one, they were unable to detect it. Spotted hyenas can also detect live prey by olfaction when they are downwind of it, although at shorter distances than for a dead animal. The mean detection distance of a live prey's odor plume was 1.1 km, with the maximum 2.8 km (Mills 1990). One potential reason why recently killed animals can be detected at such great distances is that blood has a unique odor, and scavengers are often attracted to its smell.

Likewise, predators find most nests through the nest's odor plumes because incubating birds seldom leave their nest so there are few depositional odorant trails to follow. By using odor plumes, duck nests can be detected from at least 10 m away by raccoons (personal observation), 25 m by skunks (Nams 1997; Jimenez 1999), and 30 m by red foxes (Sargeant, U.S. Fish and Wildlife Service, 1999; personal communication).

Procellariiform sea birds are also good at detecting the presence of prey by odor plumes. Leach's storm-petrels can detect and home in on an odor target at sea from distances of 1 to 12 km (Clark and Shah 1992). Miller (1942) reported that black-footed albatrosses (*Phoebastria nigripes*) were attracted to the smell of food from over 30 km.

Certainly, the distance over which a mammal can detect an odor plume depends in part on the size of the odorant source and how rapidly it is releasing odorants into the air. If humans are downwind, then they can smell a skunk's musk from a distance of 100 m, the rotting stench of a dead cow from 1 km, and the smoke of a large forest fire from 100 km. Sailors can often smell land before they can see it on the horizon.

Tsetse flies feed by sucking blood from mammals that inhabit the open woodlands and plains of Africa and find mammals, in part, by olfaction. Hargrove and Vale (1978) examined how the number of tsetse flies attracted by a mammal's odor plume would change as the size of the odorant source increased in mass. The scientists created an enclosed shed in Africa with fan-powered ventilation shafts through which odorants were released, and flies could be killed. They baited the shed with different numbers of live cattle and found that 500 kg of cattle attracted about 300 *Glossina pallidipes* per hour; 11,500 kg of cattle attracted 800 flies per hour (Figure 3.2).

Sidebar 3.1: How Can One Tell When a Predator Has Detected a Prey's Odor Plume?

Scientists are disadvantaged when studying how predators use olfaction because humans lack the olfactory acuity of olfactory predators. The method most commonly used by scientists to ascertain when a predator has detected the odor plume of prey is by watching for a change in the predator's behavior. Hunting dogs change their behavior immediately when they detect the odor plume of a bird. The dogs become excited; their ears perk up; they run or walk faster, turn directly into the wind, and move in a straighter line than before. It is also obvious when striped skunks first enter the odor plume of a food item. Normally, skunks forage by slowly walking back and forth sniffing the ground with their noses. When they smell an odor plume of food, they point their noses high into the air, change directions, and walk in a straight line toward the food (Nams 1991). Mills (1990) reports that when brown hyenas have detected the odor plume of a carcass or prey, they change directions and start to move upwind, sniff more frequently, and always sniff upwind. Mech (1970, p.197) described the change in wolf behavior thus:

Whichever way they are traveling, when their route crosses the wind flowing from the direction of the prey, the lead animals suddenly stop. All pack members then stand alert with eyes, ears, and nose pointed toward the prey. If the wolves are in an open area, they may then carry out a group ceremony with the animals standing nose-to-nose and wagging their tails for a few seconds. If they are in deep snow, they usually just pile up behind the leader and point toward the prey. Then they veer abruptly from their route and head directly toward the prey.

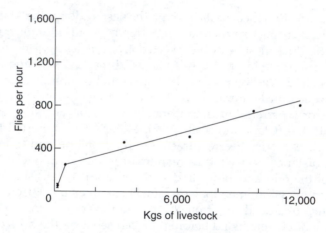

Figure 3.2 Mean number of tsetse flies (*Glossina pallidipes*) attracted hourly to a shed baited with different numbers (or mass) of cattle. (Based on data from Hargrove, J.W. and G.A. Vale, *Bull. Entomol. Res.* 68:607–612, 1978.)

Can prey reduce their odorant emission rate?

Animals that seek to hide from olfactory predators should strive to reduce the emission rate of odorants, but their ability to do this varies among odorants. For some odorants, such as those produced by anaerobic bacteria, animals have little control over their production or release into the atmosphere. At the other extreme, animals have direct control over the release of some odorants, such as sex pheromones, and can stop their emission when threatened. For other odorants, animals can slow their emission rate but not stop them entirely.

For example, deer, rabbits, woodchucks, ground squirrels, mice, and incubating birds can reduce their metabolic, respiratory, and heart rates by up to 50% when alarmed by the approach of a predator (Gabrielsen et al. 1985). By doing so, these animals can reduce their release of metabolic odorants and make it harder for an olfactory predator to detect them. For a minute or two, an animal can entirely stop the emission of metabolic odorants by holding its breath. This can be a successful strategy when an olfactory predator is within meters of its quarry but still has not located it; the predator may move on before the quarry's need to exhale becomes too great.

Animals also can reduce the emission rate of odorants that originate from their surface by lying flat on the ground, curling into a ball, and tucking limbs beneath their bodies. By grooming themselves, animals release a flood of odorants, dead skin flakes, pieces of feather, or bacteria into the air. Thus, grooming may seem to be a counterintuitive behavior for an animal that is trying to escape the detection of olfactory predators. But all the odorants and skin cells released while grooming probably would have been released during the next few days anyway. If an animal limits its grooming to times and places where it is safe from olfactory predators, then it can release surface odorants when and where the odorants will create the least harm for the animal. Hence, by grooming, an animal can minimize the number of odorants it releases when located in more dangerous places or when engaging in more dangerous behavior. In this regard, it is worth noting that grooming is often concentrated on the feet or paws. One advantage of keeping feet as odorant free as possible is to minimize the amount of depositional odorants left behind when an animal's feet strike the ground.

Although an animal can control the emission rate of many of its odorants, predators will likely ignore these odorants when hunting. Instead, predators will key in on those odorants that have a constant emission rate or ones that their quarry is unable to control (for example, odorants produced by anaerobic bacteria). Predators that seek their quarry using these involuntary odorants will be more successful than those using odorants that can be slowed or stopped by an animal when it feels threatened. Hence, the former predators are the ones most likely to survive and reproduce.

Impact of wind velocity on odorant concentration

One way for an animal to reduce its odorant concentration in the air is to remain in areas where wind velocity is high. Initially, this seems counterintuitive because if the air is moving slowly, it may seem that the animal's odorants will fall to the ground before they get very far. However, most odorants are so light that gravity has little impact on them, and they will not drop out of the air column regardless of how slow the wind is moving. What a fast wind does, however, is dilute the initial concentration of odorants in the air column. Consider a robin that is releasing a million molecules of odorants per second into the air, and this emission rate is independent of wind velocity. If air is moving past it at a rate of 1 m/second, then each cubic meter of air moving downwind of it initially contains a million molecules. If the air instead is moving at 10 m/second, then each cubic meter of air now contains 100,000 odorant molecules. As this odor cloud moves downwind, it will disperse, and the odorant will become diluted. Hence, an odor plume will be quickly diluted to undetectable levels on windy days. This principle was demonstrated with the insect *Spodoptera litura* because the distance over which male insects could detect a pheromone source decreased with wind speed (Nakamura 1976). The atmospheric forces that control wind speed and its direction are discussed in Appendix 3.

Impact of turbulence on odorant concentration

At any instant, wind speed and direction will often vary at two points even when those points are separated by only a few centimeters. This variation is a result of turbulence, which causes airborne odorants to spread out both laterally and vertically so that what started out as a narrow, highly concentrated odor plume soon becomes a wide plume.

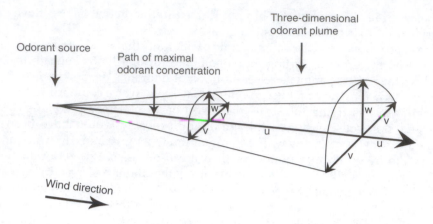

Figure 3.3 Orientation of the different axes within a three-dimensional odor plume. The *u* axis projects directly downwind of the odorant source in the direction of maximum odorant concentration and measures the stream-wise movement of odorants. The lateral distribution of odorants is recorded on the *v* axis, and the vertical distribution of odorants is measured on the *w* axis.

I should note here that although biologists usually refer to the axes of three-dimensional space as *x*, *y*, and *z*, micrometeorologists refer to axes of three-dimensional space as *u*, *v*, and *w* (Figure 3.3) with the *u* axis aligned with the horizontal direction in which the wind is moving (hereafter referred to as the *stream-wise direction*). The *v* axis is also horizontal but at right angles to the *u* axis (this direction is called the *crosswind*). The *w* axis is also at right angles to the *u* axis but is aligned vertically. It measures *updrafts* (the upward movement of air) and *downdrafts* (the downward movement of air). Throughout this book, I adopt the names and symbols used by micrometeorologists because much of this book is based on their work.

Turbulence is a measure of the variance of wind speed (σ_u, with the subscript *u* referring to the stream-wise direction) and the variance in wind direction in both the lateral (σ_v) and vertical (σ_w) planes. Although every odor plume is unique and the rate at which turbulence dilutes any single odor plume varies widely based on environmental and atmospheric conditions, it is possible to use statistics to compare many plumes to each other to discern patterns of similarity and to determine mean values. When this approach is taken, the impact of turbulence on the dispersion of an odor plume becomes more predictable and can be modeled.

One of these simple models is the Gaussian dispersion model. It assumes that, on average, the strongest concentration of odorants will be at the center of the odor plume and directly downwind from the odorant source. Hence, the *maximal odorant concentration* will be along the *u* axis, the odorant concentration in a lateral direction can be recorded on the *v* axis, and the odorant concentration in a vertical direction can be recorded on the *w* axis (Figure 3.3). The Gaussian dispersion model assumes that the actual distribution of odorants in either the lateral or vertical direction will follow a normal distribution, with the highest concentration occurring at the center of the plume (Figure 3.4). The model also assumes that σ_v and σ_w are equal to the standard deviation of the wind velocity in a lateral and vertical directions, respectively, measured during the period of time required for the odorant to reach a given point based on mean wind velocity *U*. Following the convention of micrometeorologists, *U*, *V*, and *W* refer to the mean velocity of wind along each of those respective axes over some period of time. This model has proven to be reasonably accurate, at least when there are no inversions or thermals to either stop or speed up the vertical movements of odorants (Wark and Warner 1981).

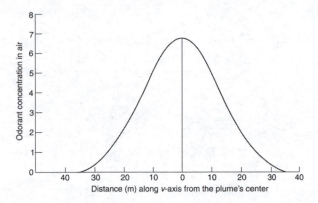

Figure 3.4 Concentration of odorants showing the lateral distribution of odorant concentrations along the *v* axis at a specific distance downwind of the odorant source. The highest odorant concentration is at the plume's center ($x = 0$).

One benefit of the Gaussian dispersion model as modified by Bossert and Wilson (1963) is that it can be used to estimate the maximum distance downwind of an odorant source in which a predator can detect the odor $U_{maximum}$ and the maximum lateral distance $V_{maximum}$ and maximum height $W_{maximum}$ that the odor can be detected. These values can be determined by knowing the rate at which odorants are released by the odorant source Q measured in molecules/second, the minimal concentration K of an odorant that a predator can detect in molecules/cubic centimeter, and the rate at which odorants diffuse in lateral (C_v) and vertical (C_w) directions. C_v and C_w are themselves directly related to σ_v and σ_w, respectively. The maximal distance directly downwind of an odorant source that a predator can detect is given by the following equation:

$$U_{maximum} = [2Q/K\pi C_v C_w u]^{1/2-n}$$

Likewise, the maximum lateral distance $V_{maximum}$ and height $W_{maximum}$ in which a predator can detect an odorant source is given by the following:

$$Y_{maximum} = [2C_y{}^2 Q/K\pi C_v C_w u e]^{1/2}$$

$$W_{maximum} = [2C_w{}^2 Q/K\pi C_v C_w u e]^{1/2}$$

For conditions of moderate winds (1 to 5 m/second) over open, level ground and no updrafts (neutral stability), Sutton (1953) reported that n equals 0.25, $C_v = 0.4$ cm$^{1/3}$, and $C_w = 0.2$ cm$^{1/3}$. Using Sutton's estimates, the equations from the Gaussian dispersion model turn simply into

$$U_{maximum} = [8Q/Ku]^{1/4}$$

$$V_{maximum} = [1.27Q/Kue]^{1/2}$$

$$W_{maximum} = [0.32Q/Kue]^{1/2}$$

Figure 3.5 A turkey vulture searching for food.

Sidebar 3.2: Using the Gaussian Plume Model to Estimate the Distance from Which a Vulture Can Detect a Carcass Using Olfaction

Smith and Paselk (1986) wanted to determine the maximum distance at which turkey vultures (*Cathartes aura*) would be able to detect a carcass by olfaction (Figure 3.5) and the authors made use of the Gaussian plume equations to do so. By measuring an increase in heart rate, the researchers determined that turkey vultures could detect butanoic acid and ethanethiol, two odorants produced by the decomposition of meat, at concentrations as low as 1×10^{-6} moles of odorant molecules/l air. When these values of K were added into the equations from the Gaussian dispersion model, Smith and Paselk were able to determine that a turkey vulture could not detect a dead carcass when the vulture was more than 3 m downwind of the carcass or more than 0.17 m above it given a wind speed of 1 m/second and an emission rate Q of 20 moles/day or 2×10^{-4} moles/second. Based on these detection distances, Smith and Paselk concluded that olfaction would not help soaring turkey vultures locate carcasses. Humans can smell ethanethiol at concentrations as low as 1×10^{-11}, much lower than vultures. Based on the same emission rates of ethanethiol and wind speed used, a person could smell the same carcass from over 100 m when directly downwind of it.

Bossert and Wilson (1963) made an interesting use of the equations from the Gaussian dispersion model to calculate the volume of air downwind of a female gypsy moth (*Lymantria dispar*) in which her sex pheromones were in a high enough concentration to be detected by a male conspecific. When the wind velocity was 1 m/second, $U_{maximum}$ was 4560 m, $V_{maximum}$ was 215 m, and $W_{maximum}$ was 108 m. On page 36, I hypothesized that detection distances should decrease with wind speed, and the work of Bossert and Wilson supports this hypothesis. They reported that, if the a wind velocity increased from 1 m/second to 5 m/second, then $U_{maximum}$ would decline from 4560 to 1820 m, $V_{maximum}$ from 215 to 97 m, and $W_{maximum}$ from 108 to 48 m (Figure 3.6).

If an animal wants to remain hidden from predators, then it should strive to keep this maximal detection length as short as possible. The Gaussian dispersion model predicts that an animal can do so by reducing its emission rate of odorants and by remaining where turbulence and wind velocities are high.

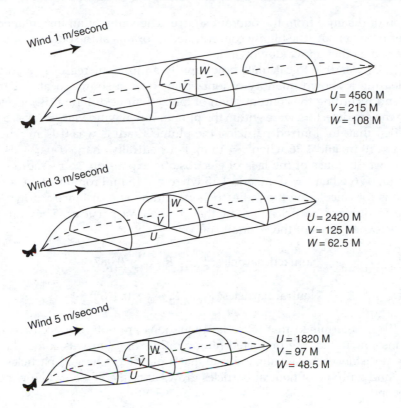

Figure 3.6 The volume of air downwind of a gypsy moth where her sex pheromones would be above detectable levels to a male conspecific at various wind velocities. Values for $U_{maximum}$, $V_{maximum}$, and $W_{maximum}$ are also provided. (Adapted from Bossert, W.H. and E.O. Wilson, *J. Theor. Biol.* 5:443–469, 1963, and used with permission from Elsevier.)

Differences in time-averaged and instantaneous views of odor plumes

Turbulence occurs at different spatial scales. Small-scale turbulence causes waves or eddies that are only millimeters or centimeters wide. Small-scale turbulence is responsible for the mixing of air within an odor plume and an expansion of the odor plumes laterally or vertically by mixing odorant-filled air from within the odor plume with odorant-free air adjacent to it. Large-scale turbulence produces eddies that are meters or kilometers in size and causes the entire odor plume to meander laterally or undulate vertically. This occurs because these eddies are larger than the odor plume, and they push around the entire plume. Large-scale turbulence might bring an odor plume to a stationary predator located downwind of the odorant source at one minute and carry it away in the next. If an odorant detector were positioned downwind from an odorant source, then there would be some periods of time when the detector showed no activity because the detector was outside the odor plume. At other times, the detector would show high activity levels because large-scale turbulence moved the odor plume over the detector. Once the detector was within the odor plume, any variations in odorant concentration that it picked up would be because of small-scale turbulence that caused odorant concentrations within the plume to vary from one place to another.

Small-scale turbulence is better assessed by taking an instantaneous view of an odor plume. Such a view ignores large-scale eddies that cause the plume to meander and undulate and rather examines only how the cross-sectional area of an odor plume varies as a function of distance from the odorant source. The value of an instantaneous view is that it better predicts how the highly concentrated odorants at the odorant source become more dispersed as they move downwind.

Jones (1983) was able to calculate how rapidly a plume increases in a cross-sectional area by releasing a known quantity of ionized air from a source located 1 m above an open field and recording the concentration of ionized air at different distances (x) downwind when the ion detectors were within the plume. When wind velocity was 5 m/second, he determined that, for ionized particles, the plume's radius was 0.44 m when $x = 5$ m, 0.83 when $x = 10$ m, and 1.36 when $x = 15$ m. For neutrally charged particles, the plume dispersed slower because of the lack of electrostatic expansion, so its radius was 0.35 m when $x = 5$ m, 0.76 when $x = 10$ m, and 1.18 when $x = 15$ m. From his data, he developed two equations for calculating the radius (R) of a plume at a distance x downwind of the source, one for ionized particles and another for neutral particles. They are given next where $R_{(x = 0)}$ is the radius of the odorant source:

$$\text{Ionized particles: } R_{(x)} = R_{(x = 0)} + 0.087x$$

$$\text{Neutral particles: } R_{(x)} = R_{(x = 0)} + 0.046x^{1.21}$$

Jones (1983) determined that these last equations are not very accurate at long distances because, at 100 m, the predicted plume radius of 12 m was somewhat too large. Still, for short distances, his results indicate that a plume of charged particles expands at a rate of 5°, and a plume of neutral particles grows at a rate of 4.5° when wind velocities are 5 m/second.

Impact of lateral and vertical turbulence on the size of instantaneous odor plumes

The volume of an instantaneous odor plume can be operationally defined as the three-dimensional volume (m³) of air within which a particular predator can detect the presence of a particular prey by identifying at least one of its odorants (that is, the space where at least one odorant the predator uses to identify prey is above detectable levels). Because most olfactory predators are terrestrial, it is also worthwhile to define the surface area (m²) of the odor plume as the ground area where a particular predator can detect prey by identifying one of its odorants. The volume of the odor plume varies for each prey based on its emission rate of odorants and for each predator depending on its olfactory acuity for the particular odorants released by that prey. Hence, the volume and surface area of odor plumes will differ widely among species of prey and predators.

Increased lateral turbulence by itself will have no impact on either the volume or the surface area of an odor plume produced by a source that is emitting odorants at a constant rate, but it will have a major impact on the geometric shape of the odor plume. Without lateral turbulence, the plume will be a long, thin line. With high rates of turbulence, its surface area will be the same size, but now its shape will be short and fan shaped (Figure 3.7). It is advantageous for prey to avoid producing long, linear odor plumes and instead to produce short, fan-shaped ones because the latter are less likely to be detected by a predator moving across the landscape.

Increased vertical turbulence by itself will have no impact on the volume of an odor plume resulting from an odorant source with a constant rate of emission, but it will have a major impact on the surface area of an odor plume. As vertical turbulence increases, the odor plume will be drawn upward, and the area on the ground where odorant

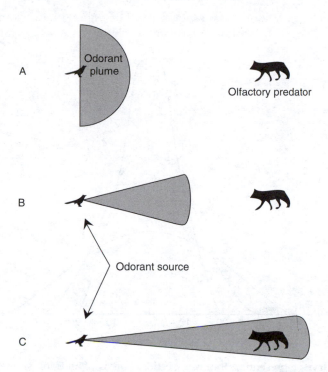

Figure 3.7 Three areas of equal size that correspond to how increased lateral turbulence affects the shape but not the area of odor plumes. (A) High rates of lateral turbulence cause the odor plume to disperse rapidly so that odorant concentrations reach undetectable levels within a short distance of the odorant source. (B) and (C) Lower rates of lateral turbulence produce long, linear odor plumes. Predators moving downwind of the odorant source are much more likely to pass through long odor plumes than short ones.

concentrations remain above detectable levels will diminish (Figure 3.8). Hence, animals seeking to avoid detection by olfactory predators should remain in areas where there are high levels of vertical turbulence.

Measurements of turbulence

In this book, I define *turbulence* as the sum of variance in wind velocity in a stream-wise, lateral, and vertical dimension (turbulence = $\sigma_u + \sigma_v + \sigma_w$) across some period of time. Hence, it is a time-averaged view of an odor plume. Many times, I refer to turbulence in only one direction. For instance, vertical turbulence is just σ_w. Turbulence measures how rapidly an olfactory plume will disperse over time. It is an excellent measure when one wants to determine the lateral and vertical spread of an odor plume at a particular distance from the odorant source (e.g., 100 m downwind of it).

Turbulence intensity is defined as the rate of turbulence corrected for the mean wind velocity U. Hence, turbulence intensity is $(\sigma_u + \sigma_v + \sigma_w)/U$; stream-wise turbulence intensity is σ_u/U; lateral turbulence intensity is σ_v/U; and vertical turbulence intensity is σ_w/U. Turbulence intensity is often inversely related to U. The reason for this is because, when it is windy, the odor plume is carried downwind quickly, and there is less time for the odor plume to expand in a lateral or vertical direction before the plume reaches a particular distance (e.g., 100 m downwind of the odorant source). Both measurements—turbulence and turbulence intensity—are important in measuring how odor plumes will disperse after they leave the odorant source. Hence, both measurements influence how easily a

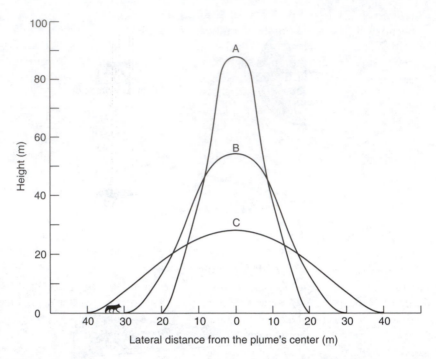

Figure 3.8 Three curves representing the cross-sectional areas of an odor plume at the same point downwind of the odorant source when there are (A) high levels, (B) moderate levels, and (C) low levels of vertical turbulence. The areas beneath the curves represent those areas where odorant concentrations are above detectable levels. Hence, if a predator was standing on the ground (height = 0) and 35 m to the side of the of the plume's center, it could not detect the odorant when there was high (A) or moderate (B) levels of vertical turbulence but could when vertical turbulence levels were low (C).

predator will be able to use olfaction to locate a prey or how likely an animal is to escape detection by an olfactory predator that is hunting somewhere downwind of it.

Spatial and temporal structure of odor plumes

The Gaussian dispersion model views an odor plume as a homogeneous cloud of odorants that differ in odorant concentration predictably based on the volume of the plume and one's location in it. In reality, an odor plume is best pictured as a floating ball of cotton. It is made up of thousands of filaments of odorants moving in the air just as clouds often appear to be made of filaments of moisture. These *odorant filaments* are separated by clean air where odorant molecules do not occur or at least where they occur in such low concentrations that they remain below detectable levels. As these filaments are moved by the wind past a stationary point, the odor plume will have a temporal element to it, a periodicity that is impacted by the width of the odorant filaments in a stream-wise direction, the width of the clean air between the filaments, and the velocity of the wind.

Hence, a passive detector that simply responds to odorants above a certain concentration (i.e., an odorant filament) will flicker on and off as odorant filaments pass by it. Within 20 m of an odorant source, the frequency at which these odorant filaments pass by a detector does not change with the detector's distance from the odorant source. But, at longer distances (over 50 m), the frequency at which odorant filaments are detected should decrease. Likewise, differences in odorant concentration between the odorant filaments and the clean air between them will diminish with increasing distance from the pheromone source (that is, the range of odorant concentrations within the plume will diminish because the odor plume is more homogeneous) (Carde 1986).

The width of odorant filaments and the width of the clean air that separates them are a product of turbulence. Turbulence itself has a temporal or structural nature that relates to the variation in the size of the eddies or waves that cause the air to flow in different directions. Waves caused by large surface features, such as hills or mountains, and large convective forces are often hundreds of meters in size. Such waves produce the large-scale eddies and cause the entire odor plume to meander or undulate. Powered by these larger eddies and embedded within them are smaller ones that are meters in length, and embedded in them are eddies only centimeters and millimeters in length. It is these small eddies that tear apart the highly concentrated stream of airborne odorants released by the odorant source into the thousands and ultimately millions of odorant filaments that compose the odor plume as it is carried farther and farther downwind from the odorant source. Hence, the energy injected into turbulence by the largest eddies is ultimately removed by viscous dissipation in the smallest eddies (Murlis et al. 1992).

One result of viewing an odor plume as a cloud of thin filaments containing high concentrations of odorants is that an individual filament may persist for a long time and be carried far away from the odorant source. Hence, an olfactory predator may catch a single whiff of odorants far beyond the distance that would be predicted based on the Gaussian dispersion model. Although much time may pass before another filament passes by, a single filament could alert an olfactory predator to the fact that there is food somewhere upwind of it and start it moving toward the food.

Effect of atmospheric instability on the vertical dispersion of odorants

Atmospheric conditions determine how rapidly odorants disperse vertically. To understand why, we need to realize that the atmosphere occurs in distinct layers and consider what causes these layers to vary in width (Figure 3.9). The *free atmosphere layer* is situated high in the atmosphere and is so called because it is free from (that is, not influenced by) forces that originate at the earth's surface. Instead, the geostrophic winds in the free atmosphere result from regional variation in atmospheric pressure and the Coriolis force

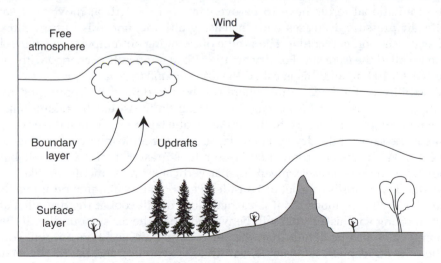

Figure 3.9 The layers of the atmosphere include the surface layer, where the atmosphere is influenced by frictional forces on the earth's surface; the boundary layer, where the forces of convection occur; and the free atmosphere layer, which is above and free from frictional and convective forces.

Table 3.1 Layers within the Earth's Atmosphere

Layer	Vertical extent	Forces influencing airflow
Free atmosphere	10–100 km	Geostrophic winds and jet stream
Boundary layer	1–10 km	Geostrophic winds and thermal turbulence
Surface layer	10–100 m	Thermal and mechanical turbulence
Olfactory zone	0.1–1 m	Thermal and mechanical turbulence
Molecular layer	1 mm	Molecular diffusion, vapor pressure

Source: Adapted from Munn, R.E., *Descriptive Micrometeorology,* Academic Press, New York, 1966.

(Sutton 1953). There is little turbulence in the free atmosphere, so an odor plume that occurs there would remain a long, thin cloud of odorant molecules that leads back to the odorant source.

Beneath the free atmosphere is the *boundary layer,* which is the part of the atmosphere that is influenced by the earth's surface. Boundary layer winds are produced from the forces exerted by the geostrophic winds above it and buoyant forces resulting from temperature variation in the air below it (Table 3.1). The air in the boundary layer is thoroughly mixed because of turbulence and thermal uplifts that occur when an air mass is warmer than the surrounding air and rises higher in the atmosphere. During the day, the boundary layer often expands severalfold over its nocturnal height because thermals are strong and reach higher in the atmosphere on sunny days (Lyons and Scott 1990).

The lower 5 to 10% of the boundary layer is called the *surface layer.* It is the part of the boundary layer that is also influenced by *mechanical friction* with the earth's surface. Turbulence in the surface layer can result either from *convective currents* or the movement of air caused created by variations in air temperature or from *mechanical turbulence* that results from frictional forces with objects on the earth's surface (hereafter called *surface features*). During the day, turbulence is mostly caused by convective turbulence; at night, mechanical turbulence prevails (Wark and Warner 1981).

The atmosphere within the boundary layer, however, tends to be *stable,* meaning that vertical air movements are suppressed, and that vertical turbulence is less than lateral turbulence. This is because more energy is required for air to move vertically than laterally because gravitational forces have to be overcome before vertical movement can occur. Normally, air pressure decreases with increasing altitude, and this in turn causes air to expand as it rises and gets colder. The profile of changing air temperatures with changing elevation is called the *lapse rate.* For dry air, the rate of temperature change with increasing altitude is 1°C/100 m, and this is called the *dry adiabatic lapse rate.*

The stability of the lower atmosphere can be ascertained by comparing the actual lapse rate (i.e., *environmental lapse rate*), which can be measured by taking temperature measurements at known heights, to the dry adiabatic lapse rate. When the environmental lapse rate is the same as the dry adiabatic lapse rate, the atmosphere is considered *neutral* (Figure 3.10). When the environmental lapse rate is greater than the dry adiabatic lapse rate, the atmosphere is *unstable,* meaning that vertical air movements are likely to occur. To understand why this is true, let us consider a gust of air containing odorants that starts to rise under such conditions. As it is carried aloft, it will cool at the dry adiabatic lapse rate, but in doing so, this air gust will be warmer than the air surrounding it. Because it is warmer, it will also be lighter than the surrounding air, and hence it will continue to rise. When the atmosphere is unstable, vertical dispersion σ_w and lateral dispersion σ_v of an odor plume will increase, and odorant concentrations will reach levels undetectable to predators in shorter distances than predicted (Figures 3.11 and 3.12).

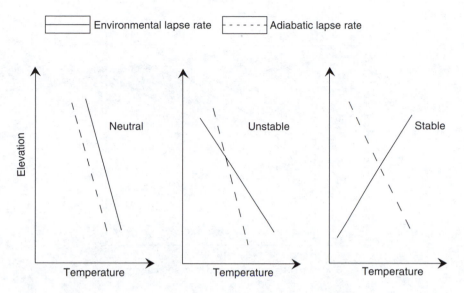

Figure 3.10 Comparison between the environmental lapse rates and the dry adiabatic lapse rates under neutral, unstable, and stable atmospheric conditions.

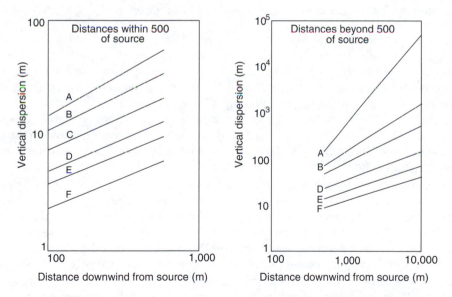

Figure 3.11 Relationship between the vertical dispersion of odorants relative to their movement downwind from the odorant source and how that relationship changes based on changes in atmospheric conditions using Pasquill's classification system: A = extremely unstable; B = moderately unstable; C = slightly unstable; D = neutral; E = slightly stable; and F = moderately stable. (Adapted from Barratt, R., *Atmospheric Dispersion Modelling: An Introduction to Practical Applications*, Earthscan Publications, Sterling, VA, 2001, and used with permission of Earthscan.)

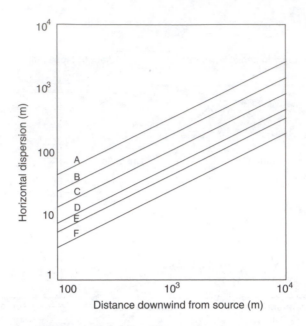

Figure 3.12 Relationship between the horizontal dispersion of an odorant relative to its movement downwind from the odorant source and how that relationship changes based on atmospheric conditions using Pasquill's classification system: A = extremely unstable; B = moderately unstable; C= slightly unstable; D = neutral; E = slightly stable; and F = moderately stable. (Adapted from Barratt, R., *Atmospheric Dispersion Modelling: An Introduction to Practical Applications*, Earthscan Publications, Sterling, VA, 2001, and used with permission of Earthscan.)

When the environmental lapse rate is lower than the dry adiabatic lapse rate, the atmosphere is stable (Figure 3.10). Under these conditions, if a puff of air containing odorants is carried aloft, it will cool at the dry adiabatic lapse rate, but when it does so, it will be colder than the air surrounding it. Because it is colder, it will also be heavier than the surrounding air; hence, the air puff will begin to sink back from where it originated (Wark and Warner 1981). When the atmosphere is strongly stable, the vertical dispersion of odorants will be suppressed, and odorant concentrations close to the ground will remain above detectable levels for greater distances than during normal conditions (Figures 3.11 and 3.12).

One of the best ways to visualize how atmospheric stability affects the dispersion of odorants is to observe the plume of smoke emitted from a chimney. When atmospheric conditions are unstable, looping plumes develop because there is a rapid vertical transport of air (Figure 3.13A). Such conditions exist mainly on sunny days when the sun heats the earth's surface. Under neutral atmospheric conditions, cone-shaped plumes develop (Figure 3.13B). These cone plumes are observed on overcast days or when it is windy. A fan-shaped plume occurs when the smoke moves laterally but not vertically and results from strongly stable atmospheric conditions that suppress the vertical movement of air (Figure 3.13C). Such conditions occur on clear, calm nights.

Diurnal changes in atmospheric stability

Atmospheric stability often changes during the course of a 24-hour period. At night, the air closest to the ground is usually the coldest and above it exists a stable layer of warm air that caps the cold air and prevents it from rising. (This is called an inversion). Under such conditions, the boundary layer is narrow, and odorants can disperse vertically up to

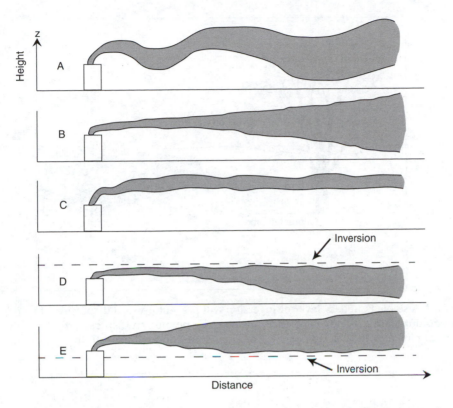

Figure 3.13 Impact of different atmospheric conditions (A = extremely unstable; B = neutral stability; C = stable; D = inversion above the odorant source; and E = inversion beneath the odorant source) on environmental lapse rates and on the typical movement and shape of smoke or odorant plumes. During stable conditions (C), the plume is vertically flattened but wide horizontally; this is difficult to depict in a figure showing a view from the side.

the inversion, but the stable boundary of the inversion prevents the dispersal of odor above it (Figure 3.13D). When the inversion is within a few meters of the ground, an odor plume cannot disperse upward but instead will hug the ground, creating ideal conditions for predators to detect an odorant source from a long distance (Figure 3.14). Sometimes, the odorant source might be above a low-lying inversion, and the odor plume cannot disperse beneath it (Figure 3.13E). Such would be the case when a bird is roosting in a tree or on top of a hill and the inversion is beneath it. When this happens, the bird's odor plume would remain above the inversion, and the bird would be undetectable to an olfactory predator located beneath the inversion (Figure 3.15).

Atmospheric conditions usually change after dawn as solar radiation begins to heat the ground surface, which heats the air immediately above it. Once the air beneath the inversion becomes warmer than the air above it, the inversion breaks down as heated air rises into the atmosphere to be replaced by cooler air, which in turn is also warmed by the earth's surface and begins to rise. This process continues during daylight hours, especially on warm, sunny days. Because warm air cannot rise at the same time and place where cold air is sinking, warm air rises in columns that are surrounded by a downdraft of cooler air (Figure 3.16). Sometimes, the winds in these rising air columns start to rotate, and when they do, their speeds can reach 3 to 4 m/second. This speed is sufficient to pick up dust or leaves, and this makes the columns of rising air visible to people and gives them their name, *dust devils* (Kaimal and Businger 1970; Fitzjarrald 1973).

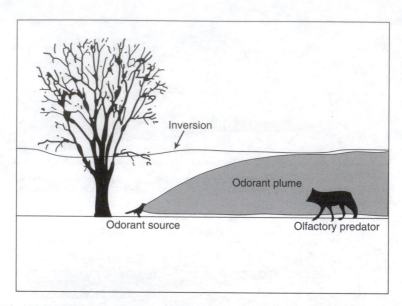

Figure 3.14 A stable boundary layer or inversion will prevent upward dispersion of an odor plume for an odorant source located beneath it. Instead, the plume can disperse only laterally, creating ideal conditions for predators to detect an odorant source located beneath the inversion.

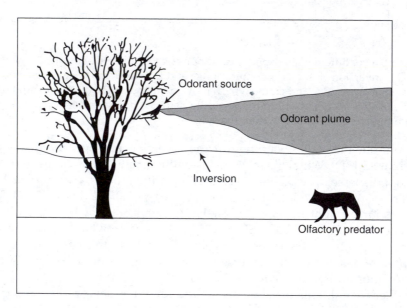

Figure 3.15 A stable boundary layer or inversion will prevent the downward dispersion of an odor plume for an odorant source located above it. Hence, olfactory predators located beneath an inversion cannot detect prey located above it.

Thermal updrafts are much bigger than dust devils and can reach as high as the free atmosphere and have a width of 1.5 times the height of the boundary layer. In the morning, thermals may only be 50 m high and 75 m wide, which includes both the area of updraft and the surrounding downdraft. By afternoon on warm, sunny days, they may be more than 2 km high and 3 km wide (Figure 3.17). Thermals can be viewed as giant conveyer belts transferring heat from the ground surface to upper parts of the boundary layer (Stull 1988).

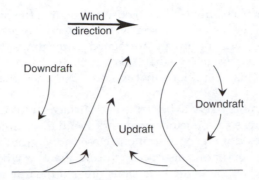

Figure 3.16 As the sun heats the air close to the ground, warm air rises in a column to be replaced by cooler air that descends from around the outside of the column.

Figure 3.17 On warm sunny days, the air close to the ground is heated and rises in giant thermals. As it rises and cools, the water within it condenses to form cumulus and cumulonimbus clouds. (Courtesy of Davis Archibald.)

Because of thermals and dust devils, airborne odorants are quickly drawn upward during sunny days until the center of the odorant's mass is half as high as the boundary layer and the odorant is dispersed throughout the boundary layer. Hence, odorants quickly dissipate to below detectable levels when there are strong thermals.

The *wind profile*, or how wind velocities in a stream-wise direction change with height, often varies considerably between day and night. During the day, especially sunny ones when thermals predominate, wind velocities vary little with height. This happens because the boundary layer is well mixed during the day so that the air in the boundary layer moves at a uniform velocity. In contrast, at night when the atmosphere is stable, wind speeds are slow when close to the ground but increase considerably with height. This results because the boundary layer is stratified at night, and only the air close to the ground is slowed by frictional forces of the earth's surface.

Impact of atmospheric instability on olfactory predators and their prey

Variations in atmospheric stability have a major impact on the ability of olfactory predators to use odor plumes to detect the presence of prey. By hunting during periods when the atmosphere is stable, olfactory predators are able to detect prey from greater distances.

Evidence for this comes from Shivik's (2002) observation that the time required for trained search-and-rescue dogs to locate a person was positively correlated with atmospheric pressure, and the latter was positively correlated with atmospheric stability. Further, Roberts and Porter (1998b) discovered that turkey nests were more likely to be depredated during warm, wet periods (conditions that would exist when the atmosphere is stable) than during cool-and-dry periods.

The atmosphere is more likely to be stable and surface winds calmer during the night than the day. For these reasons, olfactory predators spend more time hunting at night than during the day because odorant concentrations close to the ground are higher at night.

Stable atmospheric conditions also occur during the day when the sky is overcast, when wind is still, or when low-lying inversions prevent the upward spread of odorants. Hence, olfactory predators that hunt during daylight hours should concentrate their hunting activities during these periods of time. In fact, bird dogs are more successful in locating game under such weather conditions. For these same reasons, prey should seek refuge in areas where they are safe from olfactory predators (such as in trees where they are off the ground) during the night or when these conditions exist during the day. If prey must forage in areas where they are vulnerable to olfactory predators, then they should do so, to the extent possible, when the atmosphere is unstable (for example, during the day or when strong thermals exist).

Pasquill (1961) developed a simple system for measuring atmospheric stability that requires no equipment. His system is explained in Appendix 4. I encourage biologists to use his system to measure atmospheric stability whenever they are conducting an experiment or collecting data that might be influenced by atmospheric conditions.

chapter four

Using odor plumes to locate prey and the impact of convection

Locating prey through airborne odorants

To be successful, an olfactory predator not only must detect the presence of prey but also must be able to determine its location. Following an odor plume, however, is not as easy as following a depositional odor trail. For one thing, it is more difficult for a predator to track an odor plume than the depositional odor trail because an odor plume lacks sharp gradients. If a predator strays from a depositional odor trail, then it can quickly determine its error because it will detect a rapid decrease in odorant concentration. In contrast, a predator following an aerial odor trail is immersed in a cloud of odorants, making it hard to tell where the odorant concentration is greatest. At least for insects, it is much easier to detect an odor plume than to follow it to its source. For instance, many male gypsy moths can detect the sexual pheromone of a female at distances of over 100 m, but few males at such distances can locate the female (David et al. 1983; Elkinton et al. 1987). Nevertheless, both insects and mammals can detect odor plumes over considerable distances and follow them to their sources.

Potential methods animals can use to locate an odor source

There are several techniques that animals employ to follow odor plumes to the source. One technique used when odorant concentrations are considerably above detectable levels and the predator is in the middle of an odor plume is to move in the direction of the greatest odorant concentration. The Gaussian dispersion model indicates that the strongest concentration of odorants should lie directly downwind of the odorant source at the center of the odor plume. Furthermore, odorant concentrations should increase as the odor plume decreases in cross-sectional area, and this in turn will occur as the distance to the odorant source decreases. Because odorant concentrations increase with decreasing distance to the odorant source, the predator can locate the source by always moving in the direction of increasing odorant concentrations. Hence, by following the line of maximal odor concentration and moving in the direction of higher concentrations, the predator should be able to reach the odorant source (for example, the prey). Furthermore, as the predator moves closer to its quarry, different odorants will begin to reach detectable levels, so it will have more olfactory information to utilize in its search.

The second method is based on the fact that, over open ground, the wind passing by any odorant source continues in a straight line for several meters, even though subsequent

---

Sidebar 4.1: Using Olfaction to Locate a Missing Trap

David Maehr from the University of Kentucky was using Sherman live traps (aluminum boxes with trapdoors) baited with oats and peanut butter to catch mice in Florida. One day a trap was missing, and there was a faint odor of skunk. He guessed that a skunk might have squeezed its head in the live trap and gotten caught. Wanting to get his trap back and to help the skunk, Maher began walking in circles away from the trap site and noted that the smell of skunk musk waxed and waned. When the smell grew faint, he changed direction until the odor was constant and then walked upwind. When it grew faint again, he changed direction until again the odor was strong and again walked upwind. In this manner, he found the skunk hiding at the base of a rosemary shrub with his tail caught in the trap. The story has a happy ending, as Maehr was able to release the skunk and retrieve his trap (D. Maehr, personal communication, 2006).

---

changes in wind direction at the odorant source produce a serpentine-shaped plume (David et al. 1982; Perry and Wall 1986). This suggests that one strategy to find the odorant source is not to follow the odor plume but rather to fly directly upwind whenever an odorant is detected. This is the strategy used by many species of moths when they detect the reproductive pheromone released by female moths (David et al. 1983; Murlis et al. 1992). It is also the strategy used by mammalian predators when odorant concentrations are just above the predator's level of detection because at low concentrations predators have only the ability to detect the presence or absence of an odorant but not its concentration.

To adopt this strategy, a predator should head upwind when it detects an odor plume. When it reaches a point where the odor can no longer be detected, it should either move back and forth lateral to the wind (called *casting about*) until it again reaches a place where it can detect the odor or remain still until a change in the wind direction brings the odor to it. Once the predator detects the odor again, it should again head directly upwind. Through this approach and some luck, it should ultimately reach the vicinity of the odorant source. Tsetse flies (*Glossina* spp.) locate their vertebrate hosts by using both methods—heading directly upwind and staying within the odor plume and following it to its source (Brady et al. 1995). Likewise, switching from upwind flight to following the path of the odor plume has been observed in Oriental fruit flies (*Grapholita molesta*) and gypsy moths (Murlis et al. 1992).

Hunting dogs also use both methods. When they detect an odor plume of a dead bird, hunting dogs usually head directly upwind. In some cases, however, a dog will follow a serpentine path to the bird. In the latter case, the dog appears to be following the path of the odor plume.

There are problems in using either of the two methods described to locate an odorant source. One problem is that both methods require the predator to move so that it can compare odorant concentrations in different locations before it can gain an initial fix on the prey's location. This creates a problem for the predator because by moving it increases the probability that the prey will see it and flee before it has learned the prey's location. Another problem is that neither method can be used to locate an odorant source instantaneously. Instead, both methods require sequential comparisons of odorant concentrations, and this takes time to accomplish. The delay can be considerable if the animal has the misfortune to start off by moving directly away from the odorant source.

Animals can prevent this waste of time and increase their efficiency in locating odorant sources if they can locate them without moving. Because animals have two nostrils and each samples air from slightly different parts of the odor plume, animals (including humans) can locate the direction of an odorant source by comparing the odorant concentration in the air inhaled by each nostril. Animals can also ascertain the direction to an odorant source by comparing the differences in the arrival time of an odorant at each nostril. Humans can detect differences in odorant arrival times of approximately 0.2 m/second. This level of acuity is a result of humans and animals possessing specific olfactory nerves that code information about the intermittency of odorants (Atema 1995; Christensen et al. 1996; Murlis 1997).

Humans can detect the direction of an odorant source with a precision of 7 to 10° even though human nostrils are only separated by a few centimeters (Bekesy 1964). Predators that rely on olfaction to locate prey have even finer precision in locating odorant sources. Dogs have an interesting feature for sharpening their ability to locate odorant sources by comparing olfactory stimuli reaching each nostril. They have a flap of skin on the side of each nostril that they can open when trying to locate an odorant source. When open, each nostril draws more air from the side of their nose. This allows them simultaneously to sense odorant concentrations from two parts of the odor plume that are farther apart, and this helps them to locate the direction of an odorant source with greater precision.

Of course, knowing the direction of an odorant source does not tell a predator where the source is located—only the direction to it. The predator must also determine how far away the odorant source is located. There are three techniques it can use to accomplish this task. The first is to compare odorant arrival times and concentrations between its two nostrils; by doing so, it can gain some depth perception. The second is to note contextual changes in the odor plume. As a predator approaches prey, not only will odorant concentrations increase, but also more and more odorants will reach concentrations that are detectable to the predator. Detection of such odorants could inform the predator of its distance to the prey's location. The last approach is to determine the direction of an odorant source (or prey) in the vertical plane (Figure 4.1). This technique works because most predators can elevate their noses off the ground; most of the prey they are seeking are hiding on the ground beneath their nose. As a predator approaches the prey, the direction of the odorant source will be angled downward beneath the horizon. By projecting this perceived downward angle to where it hits the ground, predators will be able to get a fix on where the quarry is hiding on the ground. The ability of a predator to use this technique to locate the quarry's position on the ground will increase in precision as the predator gets closer to the quarry and this perceived angle points in a more downward direction.

When dogs are hunting birds, they often move their heads up and down as they get closer to their quarry. This behavior helps them estimate how far ahead the bird is hiding

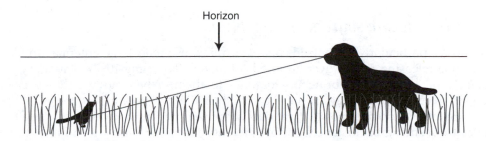

Figure 4.1 Over short distances, olfactory predators can estimate their distance to the quarry by noting the downward direction to the odorant source in the vertical plane.

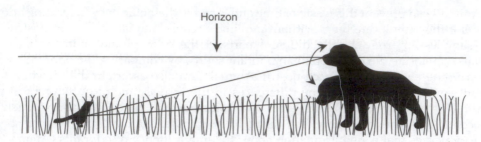

Figure 4.2 Olfactory predators can get a more accurate estimate of their distance to the quarry by moving their heads up and down and noting the direction to the quarry from different heights.

by determining the vertical angle to the bird from two different heights (Figure 4.2). Hyenas also move their heads up and down when they first enter the odor plume of a carcass or a prey (Mills 1990). I once observed a raccoon do the same thing at night when it was getting close to a nesting duck. After doing this, the raccoon rushed upwind at the incubating duck, but the duck took flight just in the nick of time. Her eggs were not so lucky.

Sidebar 4.2: Another Instinctual Response of Humans to Elude Olfactory Predators

When people are suddenly startled or scared, they instinctually gasp. This can be quite noticeable in a movie theater during a suspenseful movie, and it occurs when the monster or bad guy suddenly appears near or behind the hero or heroine. We gasp instinctually—we do it without thinking or wondering how gasping might increase our chances of survival. However, there are several benefits from this instinctive response. The most obvious benefit is that gasping fills our lungs with air, and this gives us the oxygen we might need to fight or flee. A less-obvious benefit is that by filling our lungs with air, we can hold our breath and go longer without having to exhale. Doing so stops the emission of metabolic odorants as long as we can go without breathing. If an olfactory predator is nearby but has not yet located us, then this break in the plume of metabolic odorants may be just enough for the predator to move away, hopefully far enough that we can make a dash for safety. Although most people do not realize that we hold our breath to elude a nearby olfactory predator, the instinct is observed commonly enough that we have a well-known phrase to describe it: waiting with bated breath.

How moths locate sources of odor plumes

Male moths, particularly lymantriids and saturniids, use their large antennae to locate a pheromone-releasing female over hundreds of meters (Kaissling 1997). Stationary moths must rely on the wind to bring pheromones to them. When moving, however, their antennae sweep through space at the speed of their flight plus or minus any wind speed. Unfortunately, we do not know a great deal about the specific methods moths use to find females from such great distances or how predatory insects follow odor plumes to locate prey. However, much research has focused on how male insects locate females over short distances (less than 10 m) because such questions are conducive to laboratory experiments in which variables can be closely controlled.

When male moths such as gypsy moths detect a pulse of pheromone, they fly upwind at an optimal ground speed. If the wind changes direction or speed while moths are in flight, then they might get blown off course. However, moths are able to correct for wind changes by observing their movements relative to the ground beneath them. Hence, they change the direction of their flight to correct for any change in crosswind, increase their wing beat when flying against a headwind, or slow their wing beat when helped by a tailwind (Kennedy 1983; David 1986). If too much time elapses without them detecting a new pulse of pheromone, which would happen if they leave the odor plume or encounter a large patch of clean air, then they stop heading upwind and start casting about. This increases the probability that moths that have left the odor plume will either fly into it again or remain flying in the same general area until the plume reaches them again. Once they detect a new odorant pulse, they again start flying upwind.

When moths are more distant from a pheromone source, the odor plumes become more diluted, and pheromone pulses received by the moth are weaker. Under such conditions, male moths fly longer crosswind tracks before counterturning than when pheromone pulses are strong (Willis et al. 1991). Such behavior increases the probability that they will pick up the odor plume while still at a distance to the pheromone source and to more or less hover in place when near it.

Surprisingly, upwind flights stop if males are engulfed in a homogeneous cloud of the pheromone (Kennedy et al. 1981; Willis and Baker 1984). Instead, male moths must receive distinct pulses of the pheromone to begin and maintain their upwind flight. As noted in Chapter 3, odor plumes are composed of numerous odorant filaments, and male moths detect these filaments as they pass through them. Moths can detect rapid changes in pheromone concentrations and will orient toward a pulsed odor stimulus that lasts only a few milliseconds (Kaissling 1997). The intensity of the odorant within a pulse is less important than the frequency with which these pulses are encountered by the moth. When optimal pheromone pulses exist, moths fly directly upwind without counterturning, increase their flight speed, and locate the pheromone source quickly. For many moths, upwind progress is maximized when pheromone pulses are repeated at a rate of 3/second (Kaissling 1997).

Likewise, walking moths turn and start moving upwind when they detect a pheromone pulse. If the wind stops bringing these pulses by them, then males start walking in small circles. This circular walking pattern prevents males from wandering too far and keeps them in the area where the odor plume is most likely to return (Kaissling 1997).

How tsetse flies use odor plumes to find their hosts

Tsetse flies locate their vertebrate hosts primarily by olfactory means: detecting acetone, 1-oceten-3-ol, several phenolic compounds, and CO_2 exhaled from their victim's lungs (Voskamp et al. 1998). Tsetse flies can detect the CO_2 plume of an ox when they are 90 m downwind of it (Vale 1977). Hence, unlike gypsy moths that locate small females using their pheromone odor plumes, tsetse flies follow the CO_2 plume of a much larger odorant source (for example, an elephant or a herd of buffalo). Tsetse flies seeking prey either fly about seeking an odor plume or remain still until one comes by them. In either case, on perception of the plume, they fly upwind for a considerable distance and turn back when they lose contact with it. Once close, they visually inspect potential prey, and if it is not a host, then they continue their upwind flight. The ability of tsetse flies to locate hosts using a combination of olfactory and visual cues is excellent. Over 80% of the flies that get within 100 m downwind of suitable prey actually find it (Brady et al. 1989).

Do predators develop olfactory search images of their prey?

When an animal is using its vision to search for something such as food, it often takes considerable time to locate the food the first few times, but its searching becomes more efficient with experience, and it can find food more rapidly. This occurs because animals learn the visual characteristics of their food and develop a *search image* for items with the visual characteristics of food. Likewise, animals can develop an *olfactory search image* for those odorants that betray the presence of food (Soane and Clarke 1973). This ability greatly increases the efficiency with which animals use olfaction to detect food. For instance, Nams (1991, 1997) showed that captive skunks could use smell to detect the presence of food over greater distances after prior experience searching for that specific food item. Over the course of a 7-day period during which skunks were exposed to a pseudonest each day, their detection distance for a nest increased from 2.5 to 25 m, a tenfold increase.

Impact of wind velocity on the ability of predators to locate prey using odor plumes

Predators have difficulty following odor plumes back to prey when there is no wind. Under such conditions, hyenas have difficulty finding the exact location of prey or a carcass, and hyenas that are trying to follow an odor plume sniff in all directions (Mills 1990). When there is a steady breeze, hyenas just sniff into the wind and can find carcasses easily. Bird hunters also report that their dogs have trouble finding game when there is no wind. For instance, pointers may not freeze in time but instead overrun birds and make them flush.

The reason why a lack of wind makes it difficult for predators to locate prey is because the odor plume is spherical. The prey is located somewhere within this plume, but its exact location is difficult to ascertain using just olfactory cues. Furthermore, when the air is still, the periodic breezes are likely to come from almost any direction (Griffiths and Brady 1995). Hence, if a predator were to move upwind during one of these breezes, it is as likely to be moving away from the odorant source as toward it. Plumes are more likely to *undulate* or move up and down in a vertical plane when the air is still or moving slowly. This results because the same parcel of air will remain for a longer time over a warmer or cooler ground surface. Hence, the air parcel will have more time to become warmer and rise or become colder and sink before moving across that particular ground surface.

As wind velocity increases, the odor plume changes from a sphere surrounding the odorant source to a narrower plume trailing downwind from the odorant source (Figure 4.3). Bird hunters report that their dogs can find game easier when wind speeds exceed 3 km/hour. Brady et al. (1995) noted that the ability of tsetse flies to locate an odorant source is positively related to wind velocity from 0 to 2 km/hour.

However, as wind speeds become faster and faster, odor plumes stop becoming narrower. Instead, they begin to become wider and to increase their *meandering* or sideways movement in a horizontal plane because the increasing wind velocity causes an increase in mechanical turbulence with surface features. As wind speed increases, airflow around surface features goes from laminar to turbulent (for example, surface features cause eddies as wind increases). Furthermore, as wind speed increases at the odorant source, the concentration of odorant molecules per volume of air decreases (i.e., *odorant concentration*), and this decrease in concentration reduces the total volume and length of the odor plume (Chapter 3). Hence, as wind velocities increase above a certain level, olfactory predators may have a harder time finding odorant sources. Thus, there is an optimal wind velocity for predators to follow olfactory plumes to their source, and studies have confirmed this. Nakamura (1976) reported that the optimal wind velocity for a male moth (*Spodoptera*

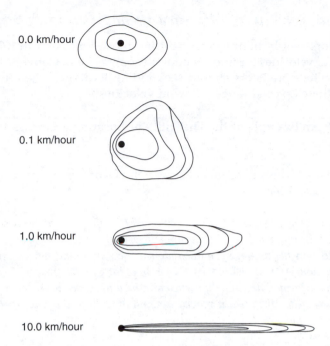

Figure 4.3 Hypothetical relationship between the shape of odor plumes and the concentration of odorants within them when there is no wind and when there is a breeze.

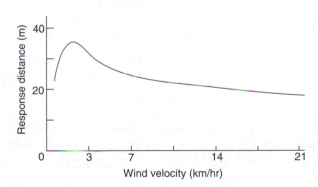

Figure 4.4 Impact of wind velocity on the maximum distance from a pheromone source that male moths (*Spodoptera litura*) can detect the pheromone plume and respond by taking flight. (Data from Nakamura, K., *Appl. Entomol. Zool.* 11:312–319, 1976.)

litura) to respond to the source of the sex pheromone released by females was 3 km/hour (Figure 4.4). Likewise, Brady et al. (1995) found that when wind speeds exceeded 3.6 km/hour the ability of tsetse flies to locate an odorant source was negatively correlated with wind speed.

Mammals also have more trouble locating an *odorant source* (that is, the object that is releasing odorants) if it becomes too windy. A survey of bird hunters that I conducted showed that dogs have difficulty finding prey when wind velocities exceed about 10 km/hour. Hence, there is an optimal range of wind speeds to locate odorant sources for pointers (3 to 10 km/hour), (unpublished data, 2006), tsetse flies (2 to 4 km/hour), and moths (2 to 4 km/hour) (Nakamura 1976; Brady et al. 1995). Unfortunately, the range of optimal wind velocities for locating olfactory sources has not been determined for most olfactory predators.

Impact of wind velocity of olfactory predators and their prey

Olfactory predators should hunt for prey during those times and in those places where their optimal wind velocities for finding odorant sources exist. Conversely, animals trying to hide from olfactory predators should seek hiding locations where the wind is either below or above these optimal ranges of wind velocities.

Sidebar 4.3: An Example of the Difficulty of Locating a Carcass When There Is No Wind

Mills (1990) made this interesting observation of a brown hyena trying to locate a carcass on a still day:

At point A, the hyena started sniffing, obviously having picked up a scent [Figure 4.5]. At point B, it started circling around in a random fashion over an area of 0.5 km². After 45 min the brown hyena abandoned the search and moved off in a north-westerly direction. After it had moved 1.4 km to point C, a southwesterly wind started blowing, and the hyena immediately turned around and moved back toward the original area. When it reached the area, it moved upwind quickly to the springbok carcass.

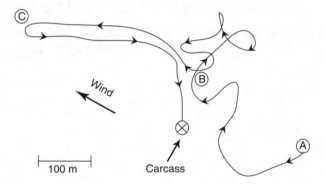

Figure 4.5 A map showing the movements of a hyena while trying to find a carcass by its odor. (Adapted from Mills, M.G.L., *Kalahari Hyaenas*, Unwin Hyman, London, 1990, and used with permission of Blackburn Press, Caldwell, NJ.)

Effect of variable wind speed and direction on use of odor plumes to locate prey

If one looks at cross sections of an odor plume at different distances from the odorant source, somewhere close to the middle of the plume is the site where the odorant concentration is highest. By connecting these sites of *maximal odorant concentration* (MOC) within each plane, one can create an aerial trail of MOC that a predator could follow if it were trying to locate the prey by always moving in the direction of the higher odorant concentrations. This MOC trail, however, is not a straight line because the wind direction changes over time at the odorant source, and it will also vary all along the odor plume. This will cause the MOC trail to curve and twist (Figure 4.6). Wind speed will also vary at the odorant source over time, and because of this, the odorant concentration will vary across time as it leaves the odorant source. During lulls, the odorant concentration will be high; during gusts, the concentration will be low.

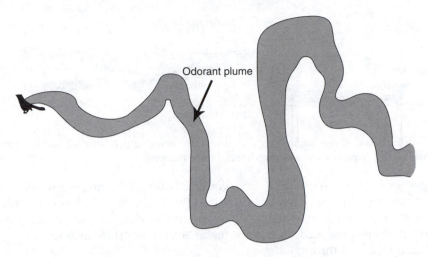

Figure 4.6 Changing wind direction across time will cause odor plumes to meander.

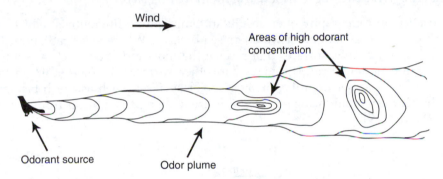

Figure 4.7 Isopleths (lines of equal concentration) of odorant concentration within an odor plume. Varying wind speeds (σ_u) at the odorant source will cause variance in odorant concentrations along the odor plume. This variance will make it difficult for predators to follow odor plumes to the odorant source.

As the odor plume moves downwind, this variance in initial odorant concentration will persist. The difficulty for olfactory predators is that they will encounter nodes of high odorant concentration along the MOC trail that are entirely surrounded by areas of lower odorant concentrations (Figure 4.7). When these are encountered, predators will have no clear indication of which way to move to remain along the MOC trail and may lose the trail entirely. Nams (1997) reports that when skunks were trying to follow MOC trails back to artificial nests, a slight change in wind direction often caused them to lose the trail.

Updrafts make it especially difficult for olfactory predators that are trying to follow odor plumes to their source. Because most olfactory predators are terrestrial, they can only sample the air for odorants to a certain height. I refer to this as their *olfactory zone*. The height is based on how high they can lift their noses and sample the air. The olfactory zone for coyotes might be 1 m, 0.7 m for foxes, 0.3 m for skunks, and 0.1 m for voles. When an updraft lifts an odor plume above a predator's olfactory zone, the prey has disappeared as far as the olfactory predator is concerned. When this happens, the predator's options are either to wait until the updraft stops and the odor trail again drops to its olfactory zone or to move on in the hope that somewhere ahead the odor plume will be closer to the ground (Figure 4.8).

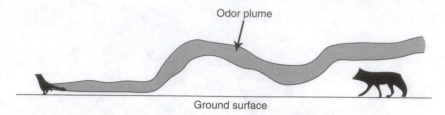

Figure 4.8 Updrafts will cause some odor plumes to rise above a predator's olfactory zone. When this happens, the odor plume becomes undetectable to the predator.

Changes in lateral, stream-wise, and vertical wind direction or turbulence are all products of the environment. They are produced by (1) differential heating of the earth's surface across time and space, which is discussed in this chapter, and (2) frictional forces to the wind that are produced by surface features that protrude into the air, which are discussed in Chapters 6 through 8.

Convective turbulence caused by local topography

Local variation in topography often results in temperature differentials, which produce local airflows, such as mountain winds or sea breezes. Where solar radiation does not strike the ground evenly, surfaces that receive the most radiation, such as south-facing slopes, become hotter during the day than other surfaces. Likewise, air above these surfaces will be warmer than the air over cooler surfaces, such as north-facing slopes. Hence, the air over south-facing surfaces will rise to be replaced by air from north-facing surfaces (Figure 4.9).

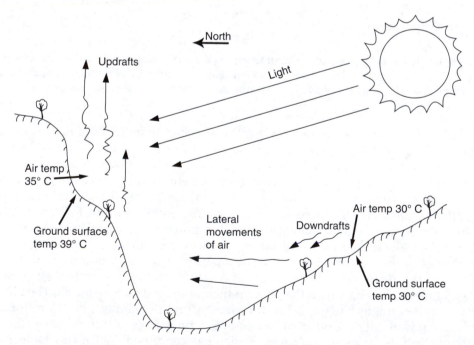

Figure 4.9 South-facing surfaces receive more sunlight than north-facing slopes, and the air above them becomes warmer and rises to be replaced by air from other, cooler surfaces.

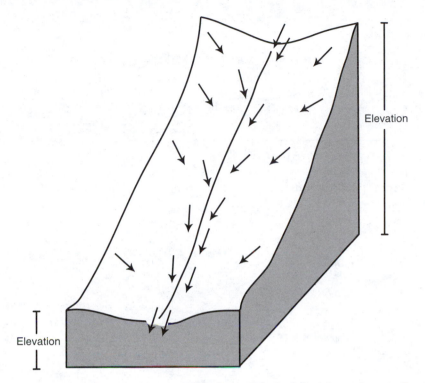

Figure 4.10 Mountain winds can occur at night when cold air aloft sinks and starts flowing down large mountain valleys or small gullies as a stream of cold air.

Mountain winds occur primarily on clear nights when cold air aloft sinks and starts flowing down mountain valleys as a stream of cold air (Figure 4.10). These winds can reach 2 to 8 m/second depending on the temperature differential of the air masses and the narrowness of the valley (Neff and King 1987). Similar air currents called *drainage winds* or slope flows can occur at night in areas where there is only a slight drop in elevation, especially if there are valleys, canyons, or ravines that the colder air can use to reach the lower elevations (Whiteman 1982; McNider and Pielke 1984). Mountain winds do not flow continuously but rather in pulses or gusts; friction with the ground can halt the downward movement of a slow column of cold air until the depth of the cold air builds to the point that its weight is heavy enough (relative to the warmer air beneath it) and is able to overcome the friction. When this happens, the cold air will burst downhill into something analogous to an avalanche of cold air.

Even gentle slopes that look flat to the eye ($\Delta h / \Delta d = 0.001$ to 0.01) can produce drainage winds of 0.5 to 3.0 m/second if the slopes continue for long distances (Stull 1988). The average slope of the Great Plains is 0.001 (rise to run), which is great enough to create drainage winds when atmospheric conditions are favorable, such as during calm, clear nights (Caughey et al. 1979). Hence, Stull (1988) argues that drainage winds can be expected over many land surfaces. The speed of drainage winds increases with increasing temperature differences, steeper changes in elevations, and longer slopes (Figure 4.11). Drainage winds extend from the ground up to a height of 10 m or more, with the fastest winds occurring 1 to 3 m above the ground (Stull 1988). When ambient mean winds are slow, drainage flows increase in velocity when they are going in the same direction of the ambient winds and decrease in velocity when heading in opposite directions. Drainage winds cease when opposed by moderate or strong ambient winds.

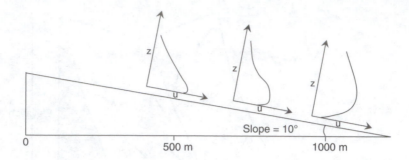

Figure 4.11 Vertical profile of a drainage wind occurring along a 10° slope at different distances down the slope.

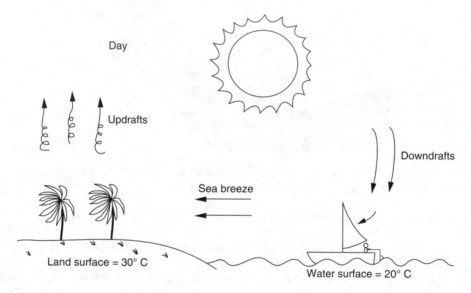

Figure 4.12 Sea breezes arise during the day as warm air over the land rises to be replaced by cooler air flowing in from the sea.

Sea breezes occur because the temperature of ocean surfaces fluctuates less during a 24-hour period than do temperatures over land. During the day, the ground heats faster than the ocean because of the greater capacity of water to absorb heat and to conduct it from the surface to deeper layers. Thus, the air over the land becomes hotter during the day than the air over the ocean, and this hot air rises to be replaced by the colder ocean air. The result is a sea breeze or wind blowing inland from the ocean during the day (Figure 4.12). The opposite pattern develops after dark because the land surfaces cool faster than ocean surfaces (Figure 4.13). That is, at night the warmer air above the ocean rises to be replaced by the colder land air, and this creates a *land breeze* or a wind that blows offshore. Oceans or other large bodies of water are not necessary to create sea breezes or land breezes; ponds, swamps, and lakes also create them. By cooling less at night, even a small pond will produce a nocturnal updraft and a flow of air from the shore into the middle of the pond or lake.

Variation in surface moisture on land surfaces can also produce the equivalent of inland sea breezes. Local air currents occur when moist surfaces are adjacent to dry ones because moist surfaces warm slower during the day and cool slower at night than drier ones (Segal et al. 1982; Yan and Anthes 1987). Such moisture gradients can occur where irrigated farmland is adjacent to a dry field or when a localized rain shower wets one area but not another.

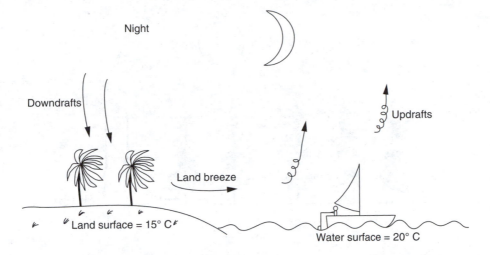

Figure 4.13 Land breezes arise during the night because the water retains heat better than land, and the air above it is warmer than over the land. Hence, at night the warmer sea air rises to be replaced by the colder air from the land.

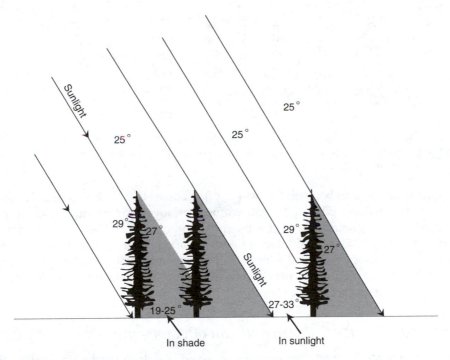

Figure 4.14 Air temperatures within a forest vary on sunny days depending on the amount of solar heating of forest canopy or ground surfaces. (Adapted from Sun, J. and L. Mahrt, *Boundary-Layer Meteorol.* 76:291–301, 1995, and used with permission of Springer Science and Business Media.)

On sunny days, temperatures within a forest are lower than outside it for three reasons: the specific heat of trees with their high water content is greater than that of soil; heat is lost through transpiration of the forest canopy; and tree crowns prevent solar radiation from reaching the ground. In fact, the light intensity reaching the forest floor can range from 2 to 40% of the light intensity outside the forest (Figure 4.14). For this reason, the

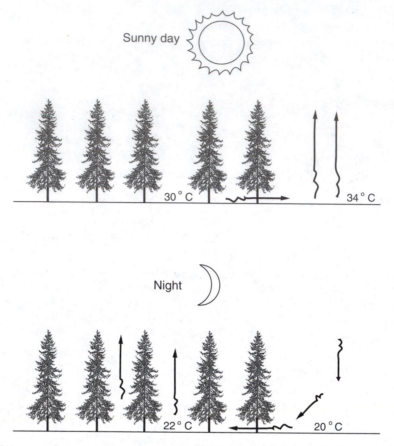

Figure 4.15 On sunny days, forest floors are cooler than adjacent open prairies, and this may produce localized airflow out of the forest. At night, the direction of airflow may be reversed because forest floors are warmer than adjacent open prairies.

summer air temperatures beneath forests are often cooler than in adjacent fields (Reifsnyder et al. 1971; Yoshino 1975). These temperature variations can produce local air currents that flow along the ground out of forests and into adjacent fields and clearings during sunny days (Figure 4.15). At night, forest floors are often warmer than open fields because the closed forest canopy retards the loss of long-wave radiation from the ground surface. Hence, after dark warm air rises from the forest floor to be replaced by colder air flowing into the forest from adjacent fields (Figure 4.15).

Impact of local convective currents on olfactory predators and their prey

Animals can use local convective currents to avoid olfactory predators. During the day and early evening, it is advantageous for prey to be on south-facing slopes or dry surfaces so that the thermals that occur there will carry their odorants up above the olfactory zone of most predators. Later in the evening, prey can move to sites where nocturnal updrafts will occur, such as wet or heavily vegetated surfaces, because these areas will stay warmer longer than other surfaces, resulting in nocturnal updrafts. Given that water bodies are warmer than ground surfaces at night, animals can hide from olfactory predators by roosting over them so that their odorants will be drawn first over the pond and then upward. Several studies indicate that both mammals and birds seek shelter over water bodies when they are particularly vulnerable to olfactory predators. For example, primates are vulnerable to predators while sleeping (Anderson 1984). In this regard, it is interesting that bonnet macaques (*Macaca radiate*), stump-tailed macaques (*Macaca arctoides*) and talapoin monkeys (*Miopithecus talapoin*) prefer to sleep on tree branches that extend over water (Gautier-Hion 1973; Estrada and Estrada 1976; Ramakrishnan and Coss 2001). Some birds also exhibit a preference for building nests on tree branches that extend over water (Quader 2006).

In the next chapter, I test one part of the olfactory concealment theory: that olfactory predators have more difficulty locating prey in areas where there are updrafts and atmospheric turbulence than where the airflow in *laminar* (i.e., smooth or straight). I do so by describing experiments that tested both domestic dogs and free-ranging predators.

chapter five

Experimental evidence that updrafts and turbulence hinder the ability of predators to find prey using olfaction

A major assumption of the olfactory concealment theory is that olfactory predators should have greater difficulty finding prey where updrafts and turbulence occur. One way to test this assumption is to create updrafts or horizontal turbulence at a specific location and then compare how quickly an olfactory predator can locate prey in such areas versus a control area where airflow is more laminar. I tested this in three experiments using domestic dogs and free-ranging predators.

Experiment 1: Do updrafts and atmospheric turbulence hinder the ability of dogs to find birds?

Methods

This experiment was carried out from March through July in both 2004 and 2005. It was conducted in horse or sheep pastures where the livestock kept the grass cropped close to the ground (vegetation height was less than 10 cm). Pastures were 1 to 2 ha in size and rectangular in shape, with the long axis running east to west. Perennial ryegrass and forbs dominated at all sites. At the start of each experiment, I randomly located a line stretching across the pasture that was perpendicular to the wind direction at the time of the test. I located two points along this line that were 20 to 30 m apart and equal distance from the center of the pasture. Treatments were randomly assigned to each of these two points.

At each point, I placed a dead bird, usually a European starling (*Sturnus vulgaris*) or the wing of a sharp-tailed grouse (*Tympanuchus phasianellus*). The dead starlings were provided to me by the U.S. Department of Agriculture's Wildlife Services (WS). The wings were provided by hunters. For each test, the same type of dead bird or wing was always placed at both sites. The birds or wings had been frozen within a few hours of the birds' death and then thawed on the day of the trial. At each site, I partially hid the dead starling or wing by placing it under some dried grass that had been cut a year before and allowed to air out in an open barn so that the grass itself would produce little odor. Similar clumps of dried grass were scattered throughout the pastures.

This experiment compared three treatments. For the vertical turbulence treatment, I placed a 0.6 × 0.6 m electric fan 0.5 m immediately above one of the dead birds and held the fan in place with wooden stakes. The fan was placed so that it was facing upward. When turned on, the fan created a vertical updraft of 1 m/second when measured at a height of 1 m above the fan. Hence, a volume of air equivalent to 0.36 m³ was displaced by the fan every second, with the updraft centered directly above the dead bird. For the horizontal turbulence treatment, I placed an identical fan on the ground 0.5 m from the dead bird and faced it so that the airflow from this fan would create a horizontal stream of air that would flow directly over the dead bird. This fan was positioned so that its breeze was perpendicular to the direction of the ambient wind at the time of the test. The third treatment was the control. For it, I used an identical fan as for the horizontal and vertical turbulence treatments, and the control fan was positioned in the same way either above or beside the dead bird. For the control, I blocked airflow from the fan by placing one board across the bottom of the fan and another board across the top so that when the control fan was turned on, it would create noise but no breeze.

All experiments were conducted in the afternoon (12:00 to 18:00 Mountain Daylight Time [MDT]) and when the mean ambient wind velocity was 1 to 5 m/second. During these experiments, ambient winds were quite variable (instantaneous velocities of 0 to 15 m/second). Hence, the fan at the horizontal turbulence site had the effect of changing the wind direction by almost 90° from when the ambient wind was 0 m/second to having almost no impact on wind direction when wind velocities were high. Hence, the fan at the horizontal turbulence site had a variable effect on both wind direction and velocity. Likewise, the vertical turbulence fan would create a column of air that was almost vertical when there was no wind and a column of air that was directed upward at a much lower angle when ambient wind velocity was high.

To start a test, I released a dog downwind of the two sites at the midpoint of the field so that the two sites were the same distance from the dog. The dog was then ordered to find the bird, and I allowed 5 minutes for it to do so. I recorded which bird the dog located first and the amount of time that elapsed between when the dog was released and when it located each of the two birds (detection time). If a dog failed to find a bird by the end of the test, then I recorded its time as 300 second (the length of the test). Sometimes, dogs became distracted during a test and found neither bird. When this happened, the test was abandoned, and no data were recorded.

If the olfactory concealment theory is correct, then dogs should find birds in the following order: (1) control birds, (2) horizontal turbulence birds, and (3) vertical turbulence birds. To test this, two treatments were paired together one at each point. The data were analyzed using a one-tailed paired t test to compare the two treatments tested simultaneously. I considered a statistical test significant if $p < .05$. No dog was tested twice in the same experiment.

Results

In the first experiment, I paired the vertical turbulence and control treatments together. Of the 17 dogs used in this test, 94% found the bird at the control site; only 41% found the bird at the vertical turbulence site during the 5-minute trial. Dogs found the bird at the control site after searching for a mean of 92 second (standard error [SE] = 20 seconds) and found the bird at the vertical turbulence site after 218 seconds (SE = 28). This difference was statistically significant ($t = 3.51$, degrees of freedom [df] = 16, $p = .001$).

When I paired the horizontal turbulence and control treatments together, 94% of 16 dogs found the control bird, but only 33% found the bird at the horizontal turbulence site. Dogs located the bird at the control site sooner (mean = 100 seconds, SE = 21) than the bird at the horizontal turbulence site (mean = 241 seconds, SE = 27). This difference was statistically significant ($t = 3.71$, $df = 15$, $p = .001$).

Fifteen dogs were tested when the horizontal turbulence and vertical turbulence treatments were paired together, and 80% found the bird at the horizontal turbulence site; 67% found the bird at the vertical turbulence site. Dogs located horizontal turbulence birds in 121 seconds (SE = 28) and vertical turbulence birds in 221 seconds (SE = 30). This difference was statistically significant ($t = 1.96$, $df = 14$, $p = .04$).

Discussion

Results of this experiment demonstrated that the ability of dogs to locate dead birds was diminished by horizontal and vertical turbulence and updrafts. The results also showed that vertical turbulence impaired the dogs' ability more than horizontal turbulence.

Experiment 2: Are nest predation rates by free-ranging predators lower in areas where updrafts occur?

The olfactory concealment theory predicts that nests will be safer from olfactory predators in areas where updrafts occur, especially at night when olfactory predators are most active. One place where updrafts consistently occur because of differential heating of the ground is along south-facing slopes because they receive more direct sunlight than north-facing slopes. This experiment examined whether predation rates on nests located on south-facing slopes are less than those located on north-facing slopes.

Methods

This experiment was conducted in Cache Valley, Utah, which is bordered on the eastern side by the Wind River Range of the Wasatch Mountains. At the base of the mountains, numerous canyons run east to west. I located 14 sites in 2 narrow canyons (Green Canyon and Dry Canyon) that had steep (approximately 45°) north-facing and south-facing slopes on opposite sides of the canyons and where both slopes had similar topography. No two sites were within 1 km of another.

The experiment was conducted in May and June 2005. Northern Utah is arid and normally sunny during these months. I assumed that the south-facing slope would be warmer than the north-facing slope during the day and stay warmer during at least the early evening. To test this assumption, I placed thermometers at both the south-facing and north-facing artificial nests at 5 of the 14 sites and checked the temperature hourly from 16:00 to 02:00 MDT. Thermometers were kept in the shade and 10 cm above the ground.

To start the experiment, I placed at each of the 14 sites one artificial nest on the north-facing and one on the south-facing slope. Both artificial nests were located at the same height from the canyon floor and usually 10 to 50 m above it. The artificial nest was made by collecting dried vegetation from the area to make a 10-cm nest bowl. Beneath this nest, I placed a small digital clock that counted hours and days and to which a metal loop was attached at the switch, which turned on the clock. A small brown chicken egg was placed on the metal loop so that the clock would not start until the egg was removed. Beside the chicken egg, I placed a dead starling that had been provided to me by WS. The birds had been kept frozen until the morning of the experiment. Dead starlings were used to provide more odor to the artificial nest.

All artificial nests were checked once every 5 days for a 20-day period. A nest was considered depredated if the egg was missing, had been eaten, or had a hole pecked through the shell because any of these would have been lethal to a developing embryo.

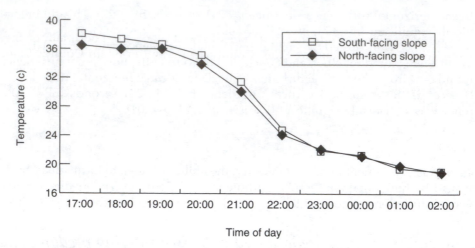

Figure 5.1 Ambient temperatures 10 cm above artificial nests located on south-facing and north-facing slopes.

When a nest was depredated, I checked the clock to determine how many hours ago the egg had been moved. By subtracting the elapsed time on the clock from the current time, I could determine the exact time the nest was depredated. I assumed that nests along south-facing slopes would survive longer than those along the north-facing slopes. Hence, I used a one-sided paired *t* test to analyze the data.

Results

From 16:00 when I started my temperature measurements until 22:00, the air above the pseudonest on the south-facing slope was 1–2°C warmer than the paired pseudonest on the north-facing slope (Figure 5.1). South-facing artificial nests survived an average of 13.7 days (SE = 2.0); north-facing artificial nests survived an average of 9.8 days (SE = 1.7). This difference was statistically significant based on a one-tailed paired *t* test ($t = 2.35$, *df* = 13, $p = .02$). Half of the nests were depredated during the hours of darkness. Six south-facing and two north-facing artificial nests survived the entire 20-day period.

Discussion

In this experiment, south-facing slopes were warmer both during the day and early evening than north-facing slopes. Hence, updrafts should occur along these slopes. The olfactory concealment theory predicted that artificial nests on south-facing slopes should survive longer than those on north-facing slopes. The results of this experiment supported this prediction.

Experiment 3: Do updrafts and turbulence hinder the ability of free-ranging predators to find artificial nests?

In the last experiment, I examined whether updrafts created by convective currents hinder the ability of free-ranging predators to locate artificial nests. Isolated surface features also create vertical and horizontal turbulence in their vicinity. In this experiment, I tested whether the turbulence created by small surface features hindered the ability of olfactory predators to locate nests.

Methods

These experiments were conducted at 41 sites located along the floodplain of the Bear River in Cache County, Utah; Box Elder County, Utah; and Franklin County, Idaho. All sites in Cache and Franklin counties consisted of open fields or pastures and were dominated by perennial grasses and forbs. All sites in Box Elder County were located either on bare mudflats or mudflats dominated by salt-tolerant vegetation less than 10 cm in height.

At each site, I located three points that were each 100 to 150 m apart, and I built a 10- to 15-cm diameter artificial nest by gathering dried vegetation from the area and sculpting it into a circular shape. I then randomly selected two of these artificial nests and placed at each a 185-l plastic garbage can made by the Rubbermaid Company. Prior to the experiment, each garbage can was left outside for 60 days so that it would weather and lose scent. I positioned the garbage can 1 m downwind from the artificial nests and placed rocks inside the garbage can to hold it in place. At one artificial nest in each site, I placed the garbage can on its side so that air passing by the can would have to rise to do so (Figure 5.2). Hence, this garbage can created vertical turbulence. At a second artificial nest, I placed the garbage can in an upright position so that air passing by it would have to curve to the side to get around it. Hence, this garbage can created horizontal turbulence. The third artificial nest at each site was left alone as a control.

At 1 week after the creation of the artificial nests and the placement of the garbage cans next to them, I wiped small brown chicken eggs several times against a dead adult chicken to provide the eggs with more odor. I then placed one of these eggs beneath each artificial nest so that the nest obscured it from view. The intention of both of these acts was to increase the egg's vulnerability to olfactory predators and reduce its vulnerability to visual

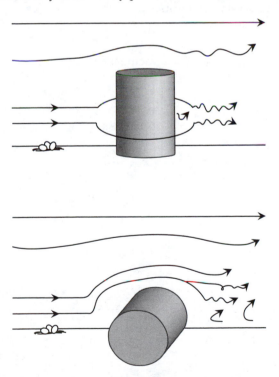

Figure 5.2 Patterns of airflow created at artificial nests where a garbage can was positioned upright or on its side.

ones. A depredated egg was defined as one that was missing, eaten, or had a hole punched through the shell.

For the first 24-hour period, each nest was checked at dawn and dusk. I assumed that if an egg were depredated at night, then it was probably taken by an olfactory predator. I also assumed that if an egg were depredated during the day, then it was probably taken by a visual predator. At 4 days after an egg was placed at each pseudonest, I rechecked each pseudonest to determine if the egg was undamaged or had been depredated. I used the Fisher exact probability test to determine if treatments changed the proportion of eggs that were depredated.

Results

Significantly fewer ($p = .04$) eggs at vertical turbulence sites (8) were depredated than at control sites (16). Likewise, significantly fewer ($p = .04$) eggs at horizontal turbulence sites (8) were depredated than at control sites (16) during the 4-day test. Predation rates did not differ between horizontal and vertical turbulence sites. Most (67%) eggs were taken at night, presumably by olfactory predators.

Discussion

These results indicate that physical features in the environment that create either vertical or horizontal turbulence can help protect bird nests from olfactory predators. The next three chapters examine the environmental features that produce updrafts and atmospheric turbulence. Once these features are identified, I assess in Chapters 10 through 13 whether young mammals and nesting birds that are vulnerable to olfactory predators use these features to hide from olfactory predators.

chapter six

Turbulence caused by isolated surface features

As a breeze moves over an isolated hill that rises above the rest of the landscape, wind velocity in the boundary layer is modified by the additional resistance to airflow created by the hill. Not surprisingly, wind speeds on the leeward side of the hill are diminished relative to wind velocities elsewhere in the boundary layer. Less obvious is that wind velocities on the windward side of a hill are also slower. This is because the hill blocks the column of air immediately in front of it.

This blockage also produces higher wind speeds at the top of the hill because the blockage increases air pressure on the windward side of the hill; when air does get around the blockage and passes over the hilltop, it accelerates to speeds higher than those that occur elsewhere in the boundary layer. This is best visualized by examining what happens when surface features such as boulders protrude into a river, causing rapids. The boulders block the flow of water as it approaches them, causing the water to back up on the upstream side of the boulders; this causes a rise in the river level. The weight of this higher water mass then pushes the water around the boulders with greater force. This is why water flows through rapids at higher velocities than it flows elsewhere in the river. This phenomenon of water accelerating through rapids is well-known and is why the areas where surface features obstruct water flow are called rapids.

The same thing happens when a column of air is blocked by a tall surface feature. Taylor et al. (1987) measured the impact of an isolated hill on wind velocities on Australia's Askervein Hill, which has a height of 116 m above the surrounding land (Figure 6.1). How much the wind increases at the top of a hill ($\Delta S = [u_{(\text{at top of hill})} - u_{(\text{normal})}]/u_{(\text{normal})}$) is related to the height of the hill h and its length L in the stream-wise direction (Figure 6.2) by the following formulas (Taylor and Lee 1984):

$$\Delta S_{\text{maximum}} = 1.6(h/L) \text{ for three-dimensional symmetrical hills}$$

$$\Delta S_{\text{maximum}} = 2(h/L) \text{ for two-dimensional ridges}$$

$$\Delta S_{\text{maximum}} = -2(h/L) \text{ for two-dimensional valleys with a depth of } -h$$

These estimates are only valid when h/L is less than 0.5 or the hill's slope is less than 33°. This means that $\Delta S_{\text{maximum}} = 1.0$ or a doubling of the normal wind speed is probably the upward bound that can be expected at the top of an isolated surface feature (Taylor and Lee 1984).

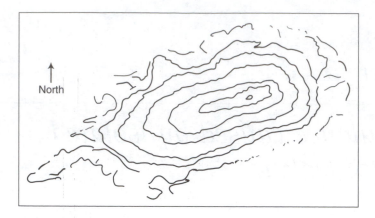

Figure 6.1 Contour map of Askervein Hill in Australia (each contour line equals 20 m).

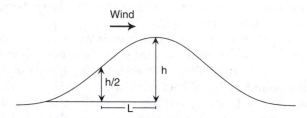

Figure 6.2 Definitions of h and L for an isolated hill. L is the horizontal distance between the hill's peak and the middle of the hill based on its height ($h/2$). L is always measured in a stream-wise direction (i.e., the same direction the wind is blowing).

Based on this equation, the proportional increase in wind speed ($\Delta S_{maximum}$) should be 0.86 or 86% faster at the top of Askervein Hill (Figure 6.1) (where $h = 116$ m and $L = 215$ m) than elsewhere in the boundary layer. This predicted value is similar to what has been actually measured (Figure 6.3).

Of course, the enhanced velocity of wind at hilltops depends on the height at which the wind is measured above the ground. For gentle hills such as Askervein Hill, ΔS at the hilltop reaches its maximal level 2 to 5 m above the ground (Taylor et al. 1987).

Mechanical turbulence caused by isolated surface features

When water flows through rapids, not only does it accelerate but also it becomes turbulent. The same thing happens with airflow; when it is blocked by a surface feature, the wind not only increases in velocity but also becomes more turbulent. The amount of turbulence produced by a surface feature will depend on its size and steepness (turbulence is especially pronounced when its slopes are greater than 20°). This turbulence is in the stream-wise (σ_u), lateral (σ_v), and vertical (σ_w) directions. The region of turbulence around an isolated feature is usually at least twice as high as the feature itself and extends downwind for a distance of five to ten times the feature's height (Lyons and Scott 1990). Measurements by Bradley (1980) at the top of Black Mountain in Australia showed that σ_u, σ_v, and σ_w were all higher there than elsewhere in the boundary layer when turbulence was measured within 25 m of the ground. When measured at a height of 87 m above the ground, σ_w still remained high, but σ_u and σ_v had returned to normal levels (Figure 6.4).

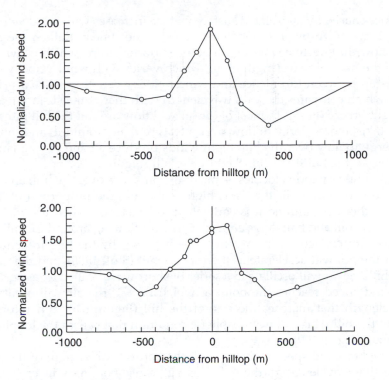

Figure 6.3 How wind velocities increase over Askervein Hill (Figure 6.1) on two different days relative to wind velocities measured elsewhere in the boundary layer. (Adapted from Taylor, P.A., P.J. Mason, and E.F. Bradley, *Boundary-Layer Meteorol.* 39:107–132, 1987, and used with permission of Springer Science and Business Media.)

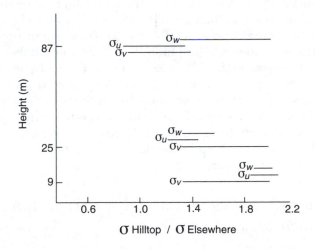

Figure 6.4 Measurements of stream-wise (σ_u), lateral (σ_v), and vertical (σ_w) turbulence at different distances above the top of Black Mountain relative to their values measured elsewhere in the boundary layer. (Adapted from Bradley, E.F., *Q. J. R. Meteorol. Soc.* 106:101–123, 1980, and used with permission of the Royal Meteorological Society.)

Turbulence caused by an isolated surface feature increases with wind velocity, but the relationship between turbulence and wind speed is not linear. Rather, the relationship depends on how the effective wavelength of the surface feature, which is twice its length, compares to the natural wavelength of air, which varies with wind velocity (Stull 1988). The relationship between turbulence and wind velocity can be predicted using *Froude's number* (Fr), which compares the air's wavelength with that of the surface feature. When there is a slight breeze (Fr < 0.1), most of the air will flow around the hill rather than over it because it takes more energy to flow over a feature than around it, and most of the air lacks the energy to flow over the hill. Light winds produce a band of stagnant air immediately in front of the hill that is blocked by the hill (Figure 6.5).

At slightly higher wind velocities, some of the air flows over the hill, and some flows around it. For a column of air that is as high as the hill, the proportion of it that flows over the hill rather than around it is equivalent to Fr. Hence, when Fr = 0.5, half of the air flows over the hill and half flows around it. The air flowing around the hill has a linear flow and does not change much in velocity as it passes by the hill. In contrast, the air passing over the hill will accelerate as it crests the hill (Stull 1988), and then once downwind of the hill, the air will oscillate in a series of short waves called *lee waves* (Figure 6.5).

When wind speed reaches the point at which Fr = 1, the air will oscillate strongly with a wavelength that matches the size of the hill (Figure 6.5), and rotors will form downwind of the hill. This is the atmospheric condition that causes the looping of smoke from a chimney (Figure 3.9).

At even higher wind speeds when Fr > 1, the natural wavelength of the air is longer than the effective wavelength of the hill. This will cause boundary layer separation and produce both a turbulent wake and a wind shadow to form downwind of the hill (Figure 6.5). Immediately downwind of the hill, the turbulent wake will be the same size as the hill. As the wind moves farther downwind, the length of the wake will continue to increase in size and decrease in intensity until the flow finally smoothes out at some distance far downwind of the hill (Tampieri 1987; Taylor et al. 1987).

Inversions will influence not only how air flows around a surface feature, but also its speed, with much depending on the height of the inversion relative to the height of the hill. When the inversion is lower than the hill (Figure 6.6A and 6.6B), all of the air must flow around the hill rather than over it, and this creates a downwind area where horizontal turbulence and vortices predominate (Brighton 1978; Hunt 1980). If the inversion is higher than the hill but still low enough to be affected by it, then a hydraulic jump, or low-pressure area, will result immediately downwind of the hilltop (Figure 6.6C and 6.6D). It is produced because some of the air crossing the top of the hill will flow downward, and this will temporarily pull the inversion down with it, creating the hydraulic jump. Further downwind of the hill, this low-pressure area between the inversion and the downwind side of the hill will produce strong updrafts and vertical turbulence as lower levels of air are pulled upward until air pressures equalize.

Impact of turbulence caused by isolated surface features on olfactory predators and their prey

Animals can use isolated surface features such as hills to hide from olfactory predators. Given that wind velocities and turbulence increase as they flow over the top of surface features, the best place for prey to hide would be at the top of them. Air that flows around the surface feature will be beneath the air that flows over it. Hence, the odor plume of an animal located on a hilltop is elevated above the olfactory zone of an olfactory predator located downwind of the hill.

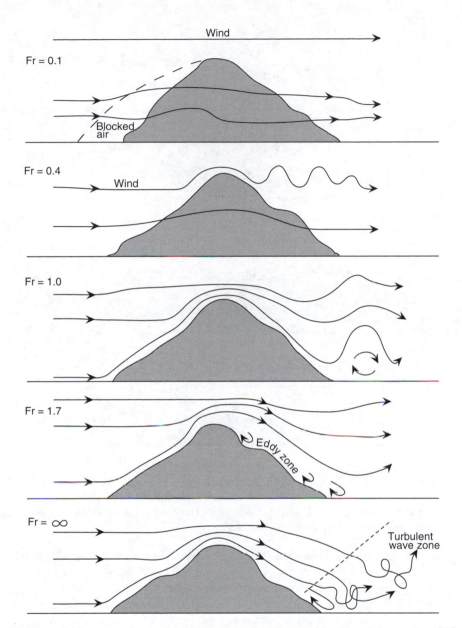

Figure 6.5 How air flows over a surface feature can be predicted by Froude's number (Fr), which compares the natural wavelength of moving air to the width of the surface feature. (Adapted from Stull, R.B., *An Introduction to Boundary Layer Meteorology,* Kluwer Academic Publishers, Dordrecht, The Netherlands, 1988, and used with permission of Springer Science and Business Media.)

Other good hiding places around isolated surface features should be directly in front of them in the area where the air is blocked by them or directly behind surface features where the turbulent wake is located. Because ΔS and turbulence increase with the steepness or slope of a surface feature, animals seeking to hide from olfactory predators should hide on steep surface features rather than on gradual ones when a choice of surface features is available.

When there are many surface features of various sizes, the best hiding place will be on or near surface features that are causing boundary layer separation, and these will vary with wind velocity. Only small features will cause boundary layer separation when airflow

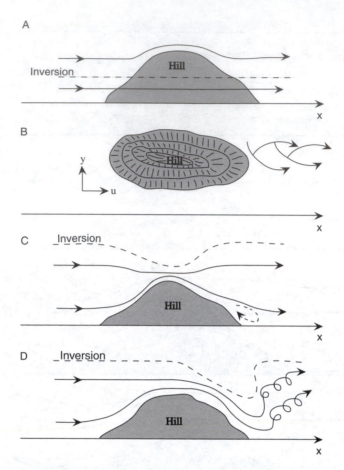

Figure 6.6 How air flows over a surface feature when there is an inversion (A) and (B) below the top of the hill or (C) and (D) above it. (Adapted from Stull, R.B., *An Introduction to Boundary Layer Meteorology,* Kluwer Academic Publishers, Dordrecht, The Netherlands, 1988, and used with permission of Springer Science and Business Media.)

is slow, but as wind velocity increases, larger features will experience boundary layer separation. The size of eddies produced by mechanical turbulence is positively correlated with the size of the surface features that produced them. Because olfactory predators are more likely to lose the scent of an odor plume where large eddies exist, the best hiding place is on top of the largest surface features in the local area that are producing boundary layer separation.

Mechanical turbulence caused by an isolated plant

A tree standing alone in a prairie causes turbulent airflow much like that of an isolated surface feature. However, the difference is that a tree is only semiporous to the wind; some airflow passes directly through the plant canopy, and some airflow goes over or around it. It is the air flowing through a semiporous surface feature that makes the turbulence caused by it to differ from solid ones. How much air flows around a tree versus through it depends on the density or porosity of the tree's canopy.

Porosity is defined as the ratio of the perforated surface area projected onto a horizontal surface to the object's total horizontal surface area. It is usually monitored when someone is interested in light penetration and is measured optically, often through photographs.

There are some obvious difficulties with using this technique to measure a three-dimensional object's porosity to airflow. One is that photos show only two-dimensional gaps that transverse the entire canopy and not the three-dimensional interstitial spaces within canopies that air uses to pass through a canopy. Another difficulty arises because barriers to the passage of airflow are not the same as barriers to light. For instance, the same amount of light will travel through a single 1-m² hole of or ten 0.1-m² holes. In contrast, more air will flow through the former than the latter because of a reduction in surface friction from the walls of the hole. Likewise, holes with smooth edges will permit more airflow than holes with sharp edges, but the smoothness of an edge has no influence on light transmission. Despite these difficulties, measuring a surface feature's porosity to light at least provides an approximation of its porosity to airflow, but the estimate will be conservative. This approximation will be increasingly inaccurate as a plant canopy increases in thickness.

The best way to measure the porosity of a plant canopy to airflow is to measure its *resistance coefficient k_r* which is defined as $k_r = \Delta P / \rho u^2$ where ΔP is the change in air pressure across the plant, ρ is the density of air, and u is the wind velocity through the plant (McNaughton 1988). Unfortunately, such data are rarely collected because they are not easy to measure. They can, however, be estimated by comparing the amount of frictional force a plant exerts to the wind with the frictional force of a solid barrier (for example, measuring the amount of force required to hold a solid board against a constant wind versus that of a porous surface of similar size and shape as the solid board). Frictional force also can be measured by comparing the difference in wind velocity in an open area to the velocity at the leeward edge of the plant.

When a concentrated plume of odorants enters the canopy of a tree, the frictional forces of the tree cause the plume to increase in both lateral width and vertical depth. However, its main effect is to cause the plume to lengthen considerably. This results because some odorants can pass through the pores of the canopy without losing any momentum and the paths of other odorants are blocked by leaves and branches, and this causes them to lose their momentum. Hence, it takes them considerably longer to pass through the canopy than other odorants (Figure 6.7).

An individual plant can also produce a net updraft or downdraft of air flowing through its canopy depending on how the plant's leaves and stems are angled. Many deciduous trees have such heavy leaves that they hang straight down from their stem and

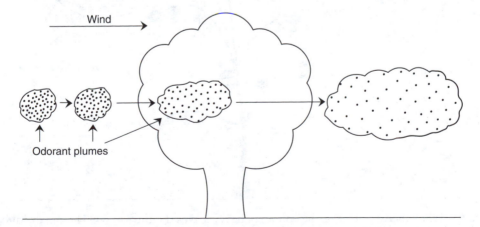

Figure 6.7 When a concentrated odor plume enters the canopy of an isolated tree, the frictional forces of the canopy cause the odor plume to increase its depth and width to some extent, but the main effect of the tree is to make the plume much longer.

Figure 6.8 Many deciduous trees have such heavy leaves that they hang straight down from their stem and are free to move with the wind.

are free to move with the wind (Figure 6.8). When there is a breeze, the bottom of the leaf will bend in the direction of the wind more than its upper edge where it is attached to the stem. Because of this, airflow will be directed downward when it passes through the canopy of a tree with this type of leaf; this will be followed by an updraft that occurs immediately downwind of the tree (Figure 6.9).

   Leaves on other deciduous trees have stronger stems attaching them to a branch. These leaves are oriented downward because of their weight but do not hang straight down because of the strength of their stems (Figure 6.10). Leaves on deciduous trees normally face outward toward the sunlight. Hence, an updraft will occur when a breeze first enters the tree's canopy because the angled leaves will deflect some of the air upward (Figure 6.11). Once the same breeze passes through to the far side of the tree's canopy and exits

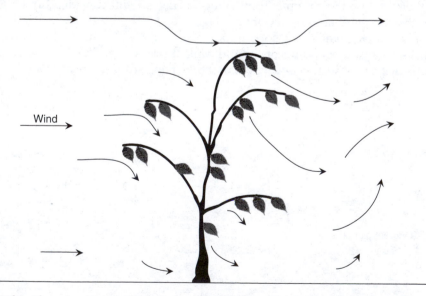

Wind

Figure 6.9 When there is a breeze, the bottom of heavy leaves (Figure 6.8) will bend in the direction of the wind more than their upper edges where they are attached to the stem. Because of this, airflow through the canopy of such trees will be directed downward, with an updraft occurring immediately downwind of the tree.

Figure 6.10 Leaves on many deciduous trees are angled downward because of their weight, but their stems are strong enough that they do not hang straight down.

the canopy, a downdraft will occur as the wind collides with the bottom surface of the leaves. The updraft will be slightly stronger than the downdraft because the latter only will occur after the breeze has passed through the tree's canopy, and hence some of the breeze's energy will have been lost to frictional forces. This pattern of updrafts and downdrafts occurring within the canopy of a single tree will create a circular airflow pattern, with the return flow occurring close to the ground below z_0 (Figure 6.11).

During winter, when deciduous trees have lost their leaves, the airflow pattern will be reversed because branches are usually angled upward from the tree trunk (Figure 6.12). A breeze entering a canopy of bare branches will likely be deflected downward from colliding with the upward-directed branches and will be deflected upward as it exits the canopy (Figure 6.13). This same pattern also exists all year for many conifers that have thin leaves projecting up from the stem (Figures 6.14 and 6.15).

Another difficulty with measuring the porosity of a plant is that it is always changing. Porosity obviously changes seasonally, especially for deciduous plants that lose their leaves in winter, but it also changes with wind velocity. As wind speed increases, air has enough energy to push leaves and branches out of its way (Figure 6.16). In doing so, it increases interstitial spaces and reduces drag, resulting in more air flowing through the

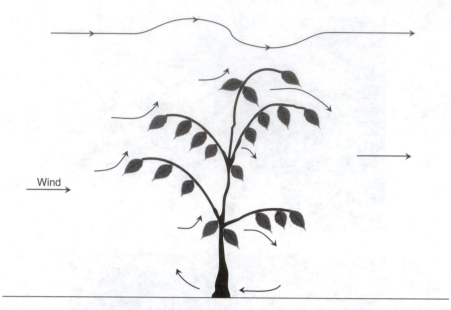

Figure 6.11 When air flows through trees with leaves that are angled downward (Figure 6.10), the air is directed upward when it first enters the tree canopy and collides with the upper surfaces of leaves and then is directed downward as it exits the canopy and hits leaves' bottom surfaces.

Figure 6.12 The branches of many deciduous trees are angled upward from the tree trunk.

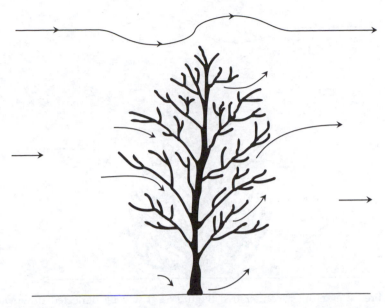

Figure 6.13 When deciduous trees have lost their leaves (Figure 6.12), a breeze entering a canopy of bare branches will first be deflected downward by colliding with the upward-directed branches and then will be deflected upward as it exits the canopy.

Figure 6.14 The branches and leaves on conifer trees project upward.

canopy. When leaves or branches are bent by the wind, they absorb kinetic energy from the stream-wise airflow. When the wind subsides and leaves, stems, and branches return to their former positions, the absorbed energy is released back into the atmosphere as turbulent kinetic energy and not into the stream-wise airflow.

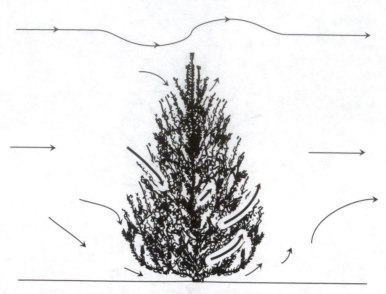

Figure 6.15 A breeze entering a conifer tree (Figure 6.14) will be deflected downward; a breeze exiting the tree will be deflected upward.

Figure 6.16 As wind speed increases, air has enough energy to push leaves and branches out of its way, and this increases the porosity of plant canopies.

Impact of turbulence caused by isolated trees on olfactory predators and their prey

Because isolated trees create localized turbulence, animals located in their immediate vicinity gain some protection from olfactory predators. When air entering a tree canopy is deflected either upward or downward, a slight updraft will result at ground level. This updraft will increase the probability that odorants from an animal located in the region of the updraft will rise above the olfactory zone of predators. Hence, the ground beneath or immediately adjacent to isolated trees is a good place for an animal to hide from an olfactory predator.

Turbulence caused by shelterbelts

From a meteorological perspective, a shelterbelt differs from an isolated tree in that its length is so long that turbulence created from wind flowing around its ends is unimportant

Figure 6.17 A shelterbelt is usually long, narrow, and positioned perpendicular to the wind direction.

and can be ignored (Figure 6.17). Instead, air can only flow either over a shelterbelt or through it. Because of this, the turbulence created by shelterbelts differs from that produced by isolated trees. Shelterbelts are designed to decrease wind velocity downwind of them, and they generally do a good job of it. Nord (1991) observed that the wind had to travel a distance greater than 15 times the height of the shelterbelt ($x = 15h$) before it regained 80% of its original speed (Figure 6.18).

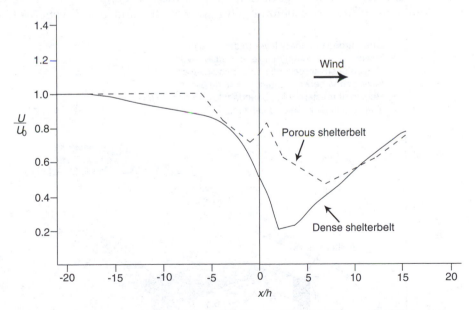

Figure 6.18 Impact of dense and porous shelterbelts on normalized wind velocities (U/U_0 or mean velocity through a shelterbelt divided by mean velocity in the open). The x axis shows distances either on the windward (negative numbers) or leeward (positive numbers) side of the shelterbelt. Distances on the x axis are reported not in meters but in units based on the height h of the shelterbelt. (Adapted from Nord, M., *Boundary-Layer Meteorol.* 54:363–385, 1991, and used with permission of Springer Science and Business Media.)

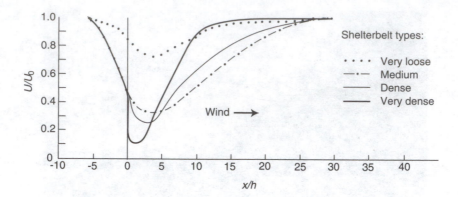

Figure 6.19 Normalized wind velocity (mean velocity through a shelterbelt/mean velocity in the open) at different distances windward (negative numbers on the x axis) and leeward (positive numbers) of shelterbelts that differed in the density of their foliage. (Adapted from Heisler, G.M. and D.R. Dewalle, *Agric. Ecosyst. Environ.* 22/23:41–69, 1988, and used with permission of Elsevier.)

Shelterbelts cause wind velocities to decrease even before the wind reaches the shelterbelt. Nord (1991) observed that speeds declined 10% even at a distance of more than twice the height of the shelterbelt on its windward side ($-2x/h$) (Figure 6.18). He also noted that within $-2h$ of a shelterbelt, the shelterbelt created an updraft as some of the air rose to flow over its top.

The shelterbelt's effect on wind velocity depends in large part on the shelterbelt's porosity (Figures 6.18 and 6.19). This is especially true in a triangular standing-eddy zone or quiet zone that extends from the bottom of the shelterbelt along the ground to a point about $5h$ leeward and then angles back to the top of the shelterbelt (Figure 6.20). Large eddies predominate in this zone behind solid shelterbelts, and the direction of mean

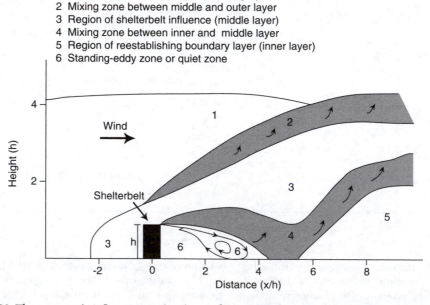

Figure 6.20 The zones of airflow around a dense shelterbelt. (Adapted from Plate, E.J., *Agric. Meteorol.* 8:203–222, 1971, and used with permission of Elsevier.)

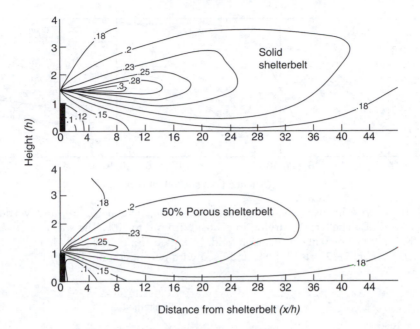

Figure 6.21 Lines of equal turbulence intensity (σ_u / U_h) leeward of shelterbelts of different porosity. Distances and heights have been standardized based on the height of the shelterbelt. (Adapted from Heisler, G.M. and D.R. Dewalle, *Agric. Ecosyst. Environ.* 22/23:41–69, 1988, and used with permission of Elsevier.)

airflow in this zone is often opposite that of the air flowing above the shelterbelt. Such eddies tend to be large, reaching to the height of the shelterbelt itself. With porous shelterbelts, airflow in the triangular zone slows but does not reverse direction (Plate 1971).

Behind and above the standing-eddy zone is a mixing region where the fast airstream flowing over the shelterbelt reunites with the slower-moving air that passed through the shelterbelt (Figure 6.20). As these two air masses mix, levels of turbulence will be high. Because the faster-moving air that passes over the shelterbelt is above the slower-moving air mass that passes through the shelterbelt, updrafts also occur in the mixing region (Figure 6.20). Not surprising, turbulence is greatest at the top of shelterbelts. It also is greater at the top of solid shelterbelts than at the top of porous ones (Figure 6.21). Shelterbelts cause an increase in stream-wise (σ_u), lateral (σ_v), and vertical (σ_w) turbulence (Figure 6.22).

When the wind is blowing at an angle to the shelterbelt instead of head on, the shelterbelt causes a directional change in the wind passing through it (Nord 1991). Such a directional change is perpendicular to the angle of the shelterbelt and has the effect of getting the airflow through the shelterbelt in the shortest distance. The result of this is that the air flowing through the shelterbelt is moving in a different direction than the air further aloft (Figure 6.23). This directional change causes lateral turbulence in the wake zone where these two air masses meet.

Impact of turbulence across shelterbelts on olfactory predators and their prey

Because of updrafts, olfactory predators will have the greatest difficulty using odor plumes to locate prey hiding in the region immediately windward of a shelterbelt ($-2h < x < 0$). The wake zone (Figure 6.20), which is $8h$ to $20h$ leeward of a shelterbelt, is a good place

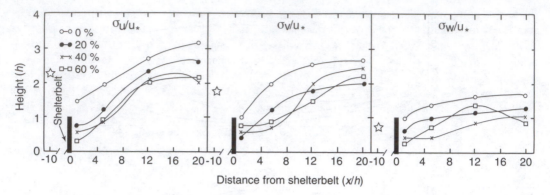

Figure 6.22 Turbulence intensity in the stream-wise (σ_u/u_*), lateral (σ_v/u_*), and vertical (σ_w/u_*) direction leeward of a shelterbelt varying in porosity from 0 to 60%. The star at $x/h = -10$ is the value for all four measurements in the open field. (Adapted from Heisler, G.M. and D.R. Dewalle, *Agric. Ecosyst. Environ.* 22/23:41–69, 1988, and used with permission of Elsevier.)

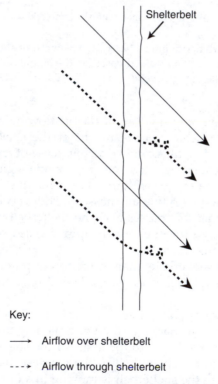

Key:

⟶   Airflow over shelterbelt

- - ->   Airflow through shelterbelt

Figure 6.23 How a shelterbelt changes the direction of wind blowing through it versus the wind direction above the shelterbelt. Air flowing through the shelterbelt experiences horizontal turbulence after it passes through the shelterbelt and is reunited with the air that passed over the shelterbelt.

for animals to hide because of its enhanced turbulence. The standing-eddy or quiet zone, which is 0 to $8h$ leeward of a shelterbelt, is not as good a place to hide as the other two zones. Animals hiding in the quiet zone, however, would gain protection from predators hunting either in the wake zone or beyond it because odorants emitted by prey in the quiet zone also have to pass through the wake zone to reach them. Hence, an animal in the quiet zone would be safer from distant predators than if it were in open field. The relative safety of these zones for animals seeking to hide from olfactory predators in order of decreasing safety should be (1) windward zone, (2) wake zone, (3) quiet zone, and (4) open field.

The widths of the quiet zone and wake zone are primarily dependent on the height of the shelterbelt. Hence, good hiding places for prey will be farther away from a tall shelterbelt than a short one. Olfactory predators should also have an easier time finding prey around a porous shelterbelt than a dense one because turbulence is positively correlated with shelterbelt density. The width of a shelterbelt or its other characteristics, such as plant composition, will be of minor importance to olfactory predators unless those characteristics change a shelterbelt's height or porosity. Olfactory predators will have an easier time finding prey around a shelterbelt when the air is blowing at right angles to it rather than at an oblique angle.

chapter seven

Turbulence over rough surfaces

The surface layer is the part of the boundary layer where the frictional forces of the earth's surfaces are felt by the atmosphere. By definition, those forces must approach zero at the top of the surface layer and must reach their maximal close to the earth's surface. Between these two extremes, wind velocities in the surface layer usually decrease logarithmically with decreasing height above the earth's surface (Figure 7.1). The steepness of the curve and how far above the ground winds will be slowed by surface friction depends in part on the roughness of the surface terrain. Air flowing over rough surfaces such as those containing boulders, trees, or hills will experience more friction than air flowing over a smooth surface (for example, a frozen lake). As a result, the surface layer above a rough surface will reach higher into the atmosphere because air currents further aloft will be slowed by the greater frictional forces exerted by the rougher surface.

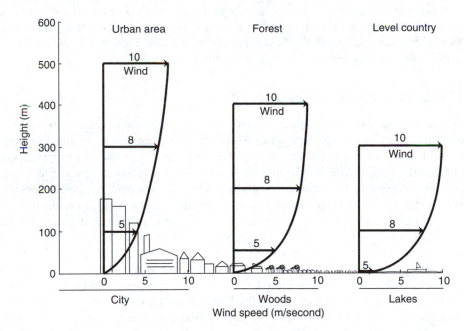

Figure 7.1 Wind velocities decrease logarithmically with decreasing height above the surface. The shape of the wind velocity profiles depends on frictional forces caused by objects on the earth's surface. In this figure, the wind above the surface layer is traveling at 10 m/second. (Adapted from Turner, D.B., *Workbook of Atmospheric Dispersion Estimates*, U.S. Department of Heath, Education, and Welfare, National Center for Air Pollution Control, Cincinnati, OH, 1967.)

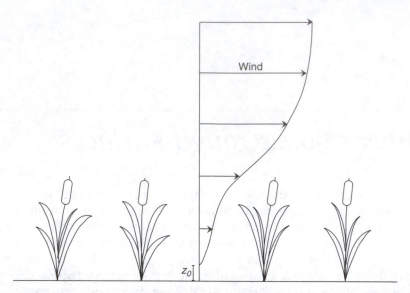

Figure 7.2 Wind velocity profiles above a cattail marsh. The length of each arrow is proportional to the wind velocity at that height. Note that wind velocity reaches zero above the water's edge.

Any time two air masses traveling at different velocities meet, turbulence will occur. Hence, turbulent eddies can be expected to occur throughout the surface layer. Within the surface layer, the faster air masses will be above slower ones (Figure 7.1). Because of this, turbulence in the surface layer will cause the lower levels of air and the odorants within them to rise and combine with the faster, higher air masses. Hence, air within the surface layer will be thoroughly mixed when flowing over rough surfaces.

Aerodynamic roughness length

An important feature of wind velocities over surfaces is that when they are measured at various heights above any surface and plotted, they do not reach zero at the surface of the ground but rather at a point above it. For instance, in Figure 7.2 the curve reaches the y axis above the origin. The height above the ground where wind speed reaches zero is also the point where eddies begin to dominate airflow. This height is called the *aerodynamic roughness length* z_0. Beneath this height, airflow in the stream-wise direction will be negligible, and it may be in the opposite direction. The aerodynamic roughness length z_0 is a characteristic of the surface and does not change with wind velocity. Its value is related to how high the surface elements protrude above the surface, with z_0 positively correlated with their height (Table 7.1). Generally, z_0 varies from 0.05 to 0.14 of the height h of the surface elements (Figure 7.3). It is about 0.05 m for tall grass pastures, 0.01 m for a mowed lawn, 0.001 m for water surfaces, and 0.0001 m for ice. Garratt (1992) suggests that $z_0 = 0.1$ m is a good estimate for most land surfaces. He also suggests that $z_0 = 0.1/h$ is a good rule of thumb, but Barratt (2001) recommends that a better one would be $z_0 = 0.03/h$.

The measure z_0 is related to the density of the elements on the surface, with z_0 reaching its highest value when there is an intermediate density of elements. For instance, it was originally believed that for a sandy surface, z_0 was 3% of the diameter of the sand grains lying on the surface (similar to Barratt's 2001 rule of thumb). However, Xian et al. (2002) demonstrated that z_0 was 13% of the diameter when sand grains were spaced farther apart from each other (similar to Garratt's 1992 recommendation that $z_0 = 0.1/h$).

Table 7.1 Measured Values of Mean Height h, Aerodynamic Roughness Length z_0, and the Zero-Plane Displacement (d) and the Ratio of d/h for Various Natural Surfaces

Surface	h (m)	z_0 (m)	d (m)	d/h	Reference
Bare soil	—	0.001	—	—	Garratt (1992)
Crops					
Wheat	0.25	0.005	—	—	Garratt (1977)
	0.4	0.015	—	—	Garratt (1977)
	1.0	0.05	—	—	Garratt (1977)
Soybeans	0.8	0.06	0.5	0.6	Baldocchi et al. (1983)
	1.1	0.08	0.7	0.6	Baldocchi et al. (1983)
Saltwater marsh (grass and sedges)					
Unmowed	0.2	0.035	0.1	0.55	Blackadar et al. (1967)
Mowed	0.05	0.01	0.0	0.0	Blackadar et al. (1967)
Savanna	8.0	0.4	4.8	0.6	Garratt (1980)
	9.5	0.9	7.1	0.75	Garratt (1980)
Forest					
Oak/hickory	23.0	—	20.7	0.9	Baldocchi and Meyers (1988)
Pine	10.5	1.0	8.0	0.76	Raynor (1971)
Pine	12.4	0.32	—	—	Hicks et al. (1975)
Pine	13.3	0.55	—	—	Thom et al. (1975)
Old-growth fir	30.0	0.12	—	—	Lee and Black (1993)
Spruce-fir	30.0	2.0	14.0	0.46	Miller et al. (1991)
Sitka spruce	6.1	0.56	4.9	0.78	Irvine et al. (1997)
Tropical	32.0	4.8	—	—	Thompson and Pinker (1975)
Tropical	35.0	2.2	29.8	0.85	Shuttleworth (1989)

Note: See Smedman-Hogstrom and Hogstrom (1978), Hicks et al. (1975), and Kondo and Yamazawa (1986) for additional data.

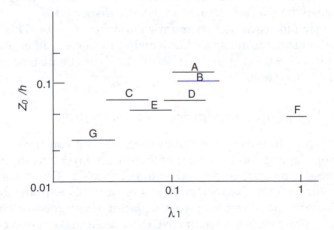

Figure 7.3 Variation of the ratio z_0/h with changing density of surface elements λ from field data using (A) and (C) trees, (B) airflow perpendicular to the rows in a vineyard, (D) and (E) wheat, (F) pine forest, and (G) airflow parallel to the rows in a vineyard. (Adapted from Garratt, J.R., *Aerodynamic Roughness and Mean Monthly Surface Stress over Australia*, CSIRO Division of Atmospheric Physics Technical Paper 29, Canberra, Australia, 1977, and used with permission of CSIRO.)

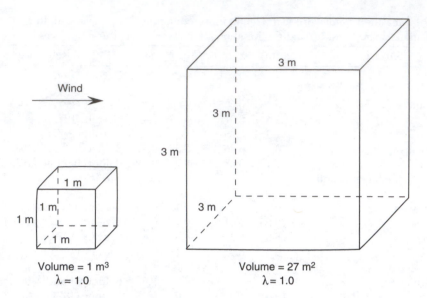

Figure 7.4 λ is defined as the cross-sectional area of the surface feature that is at a right angle to the wind divided by the ground area occupied by that surface feature. In this figure, λ = 1.0 for both square blocks because for each block the surface area that it presents to the wind is identical to the ground area it occupies.

Also, z_0 is dependent on the shape of the surface features, particularly λ, which is the cross-sectional area of the surface feature that is at a right angle to the wind divided by the ground area occupied by that surface feature (Figure 7.4). Lettau (1969) determined that $z_0 = 0.5h\lambda$ when the surface features are evenly spaced, not too closely packed together, and of similar sizes and shapes. Kondo and Yamazawa (1986) noted that z_0 can be approximated by measuring the height h and width w of each surface element i along a line of length L and by using the following equation: $z_0 = [0.25/L] \Sigma h_i w_i$. However, this equation holds true only when the surface features are widely dispersed.

The ratio z_0/h should reach its maximum value when λ = 0.4. This means that z_0/h will be maximized if elements are spaced downwind from each other at a distance of 2.5 times their height if $z_0/h = 0.1$ (Figure 7.5) (Garratt 1977) or at a distance of 8.3 times their height if $z_0/h = 0.03$ (Barratt 2001).

Impact of z_0 on olfactory predators and their prey

The shape of the odor plume and the distance an odorant can travel before it reaches nondetectable levels depends on the height of the surface layer and the amount of turbulence within it. These two variables depend on the surface's z_0. Given a choice of surfaces, animals are best hidden from olfactory predators on surfaces with the greatest z_0.

Waterfowl biologists have recommended replacing short grass or pastures with tall, rank vegetation (i.e., *dense nesting cover,* or *DNC*) because it makes good cover for upland-nesting ducks. Consequently, large areas across the Great Plains have been planted in DNC that consists of tall grasses and tall forbs. Some ducks prefer nesting in DNC (Miller 1971), and their nesting success there is higher than in unmanaged plots (McKinnon and Duncan 1999). One reason for the potential success of ducks nesting in DNC is that a mixture of plant heights would create a rougher surface and a higher z_0. This would make it harder for olfactory predators to locate duck nests in DNC than in areas where all of the plants are shorter or all of the same height, creating a lower z_0.

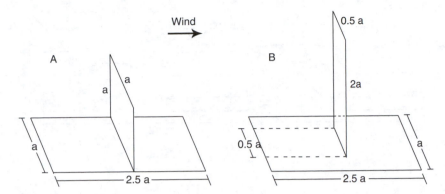

Figure 7.5 Two examples of what a ground surface would look like when the ground area around a surface element is 2.5 times the area that the surface element presents to the wind. In both (A) and (B), each surface element has a frontal area of a^2 and occupies an area on the ground of $2.5a^2$, meaning that no other surface element can occur with it. Hence, the distance between surface elements is 2.5 times the height of the surface element.

Zero-plane displacement

Bradley (1968) examined wind movements between a smooth tarmac with $z_0 = 0.002$ cm and the same surface after he constructed a field of nails 8 cm tall and in different densities. When nails were placed in an 8 × 8 cm pattern ($\lambda = 0.025$), z_0 increased to 0.20 cm. He discovered that increasing the density of nails beyond this 8 × 8 pattern did not increase z_0. The reason for this was that the wind started to skim over the top of the nails rather than travel through them as nail density increased. When this happened, the amount of friction provided by the nails decreased, and therefore z_0 decreased. The tendency for airflow to skim over the top of surface features rather than pass through them is a characteristic of the surface and is generally independent of wind velocity (Figure 7.6). This surface characteristic is called the *zero-plane displacement d* and is defined as the distance above the ground that the wind does not penetrate because it has been blocked by features on the surface.

The zero-plane displacement d is clearly related to the height of the surface features, but the actual ratio of d to h depends on the density of the surface area. For instance, when the entire surface is covered with surface elements, d has to equal their height because if there is no space between them, their tops are a solid surface. Likewise, surface elements can be spread so far apart that each becomes a solitary object that has little effect on the overall wind velocity. Hence, d and z_0 are both affected by two characteristics of the surface: h and λ. If wind speed is measured at three or more heights under neutral atmospheric conditions, then z_0 and d can be determined using a computerized regression algorithm (Stull 1988).

In most coniferous forests, d/h ranges from 0.6 to 0.9 of mean tree height (Jarvis et al. 1976). As a rule of thumb, d is often approximately two-thirds of the mean height of the surface elements for most surfaces (Figure 7.7). When tall plants are sparse, d does not occur, and Raupach et al. (1996) considered sparse canopies to be those in which the distance between plants was greater than their height. But, as the vegetation becomes thicker, the ratio d/h approaches 1. For vegetated surfaces, both z_0 and d decrease with wind velocity. This results because plant surfaces are flexible, and they will bend and orient themselves toward the direction of the wind. In doing so, their friction to the wind decreases, and because of this, z_0 and d decrease.

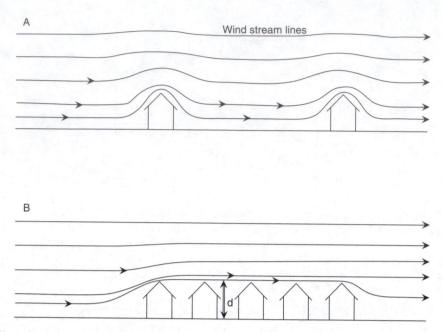

Figure 7.6 When surface features such as houses are far enough apart, each acts as an isolated surface feature (A), but as their density increases, the surface features act as a displaced surface to the wind at a height of *d*.

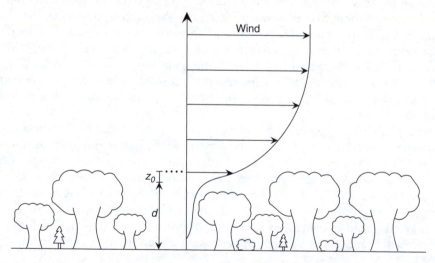

Figure 7.7 Airflow showing wind flow *u* as a function of forest height *h*. The forest canopy acts as a displaced surface to the wind with a height of *d* above the actual ground surface. The aerodynamic roughness length z_0 appears above it.

For many forests, λ is approximately half of the leaf area index of the forest canopy (i.e., the *canopy area index*, or CAI) (Raupach 1994; Raupach et al. 1996). If z_0 reaches its maximum value relative to canopy height (z_0/h) when λ = 0.4, then z_0/h would also reach its maximum value when the CAI equals 0.8 (Figure 7.8A), while *d/h* continues to increase with higher CAI (Figure 7.8B). Three conclusions can be drawn from Figure 7.8. First, turbulence will increase in forests as the CAI increases until the latter reaches 0.8. After that, the amount of turbulence will level off. Second, *d/h* decreases as CAI declines, with

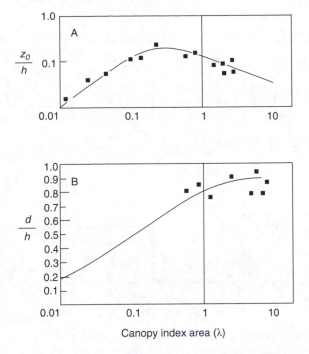

Figure 7.8 Impact of canopy area index λ on the ratio of aerodynamic roughness length to canopy height z_0/h and the ratio of zero-plane displacement to height d/h. The solid lines are the predicted relationships, and the solid squares represent actual data. (Adapted from Raupach, M.R., *Boundary-Layer Meteorol.* 71:211–216, 1994, and used with permission of Springer Science and Business Media.)

the fastest drop occurring when the CAI is between 0.05 and 0.5. Third, turbulence at the ground level will be maximized when the CAI ranges from 0.0 to 0.12 or when λ = 0.04 to 0.06.

Airflow across habitat edges

Airflow from a smooth to a rough surface

Turbulence will be especially pronounced where two habitats of different roughness come together. Consider a breeze blowing from a grassland to a forest. When the breeze reaches the forest, the air close to the ground slows because of increased resistance; air farther aloft will either continue at the same speed or actually increase in velocity (Figure 7.9). This differential change in wind velocity with height produces turbulence at the forest edge and an increase in the height of the surface layer. This turbulent flow continues for some distance downwind of the habitat edge until the airflow adjusts to the surface roughness of the forest and returns to a smooth flow. The distance downwind from the forest edge to where the turbulence ends is called the *fetch*, and its length is operationally defined as the distance between the point where the increased turbulence begins to the point downwind of the edge where only 10% of the increased turbulence caused by the forest edge still remains. A good rule of thumb is that the fetch is ten times the difference in height between the two habitat types. Where forests and prairies meet, the height will be the mean height of the forest. The area of increased turbulence will also extend upwind into the grassland because the wind begins to slow even before reaching the forest edge. As the wind blows from a smooth to a rough surface, it has to rise to get over the stalled

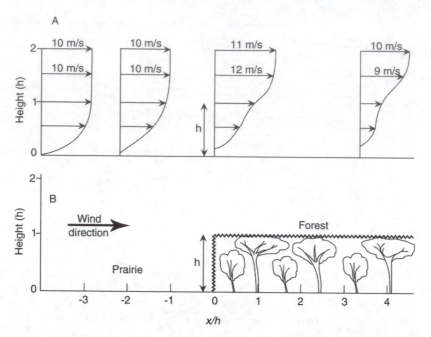

Figure 7.9 Wind velocity profiles (A) as a breeze moves from a prairie to a wood lot (B). Note that the wind accelerates to 12 m/second immediately above the ecotone.

air backed up by the taller canopy. In doing so, it will create a low-pressure area downwind of the habitat edge; this in turn will create a region downwind of the habitat edge where updrafts are strong (Stacey et al. 1994).

Irvine et al. (1997) observed that this updraft started at the forest edge and continued into the forest for a distance of over 112 m (15 times the forest height). They observed the mean wind vector in this region was not level (0°) but was instead tilted upward at an angle of 8° (Figure 7.10).

Because the forest is semiporous to the wind, not all of the air flowing past a forest edge will rise above it; some will penetrate into the forest itself. Airflow into the forest edge is particularly pronounced at the height of the tree trunks, especially if the forest lacks understory vegetation, but these wind flows do not penetrate far into the forest (Figure 7.11). Linear winds coming from the edge are prevented from penetrating far into the forest's interior because the updrafts that occur along forest edges deflect linear gusts upward into the canopy (Stacey et al. 1994). Hence, at distances from the forest edge that are greater than 15 times the canopy height ($x > 15h$), there is a dead zone that experiences less wind than other parts of the forest interior (Stacey et al. 1994; Lee 2000). This results because wind coming from the direction of the forest edge can rarely penetrate into the dead zone.

Irvine et al. (1997) observed this phenomenon across a habitat edge where moorland gave way to a Sitka spruce (*Sicea sitchensis*) forest ($h = 7.5$ m). By comparing the wind velocity of different heights over the moorland and the forest, they discovered that the wind coming from the moorland did not penetrate into the forest interior. Raynor (1971) measured wind velocity as it flowed from an open field to a pine forest 10 m tall. He found that wind penetrated into the forest for a distance of 60 m but not beyond that (Figure 7.11).

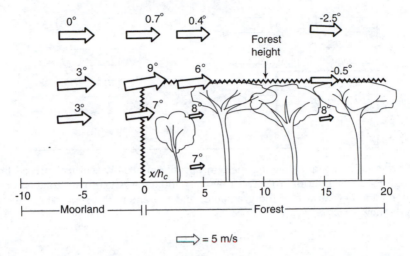

Figure 7.10 As wind flows from a moorland to a forest, updrafts are produced when the wind rises to get above the air blocked by the forest. The length of each arrow is proportional to wind velocity, and the degree sign above each arrow shows the horizontal deviation of the wind. Positive degree signs represent updrafts; negative degree signs show downdrafts. (Adapted from Irvine, M.R., B.A. Gardiner, and M.K. Hill, *Boundary-Layer Meteorol.* 84:467–496, 1997, and used with permission of Springer Science and Business Media.)

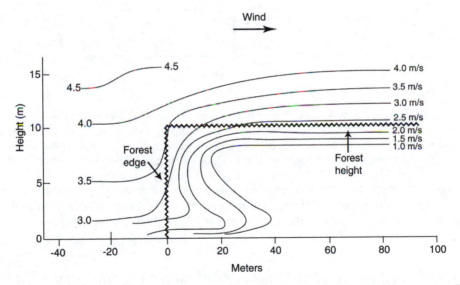

Figure 7.11 Wind velocity isopleths (lines show where the air speed is identical) for a breeze flowing from an open field to a pine forest. (Adapted from Raynor, G.S., *Forest Sci.* 17:351–363, 1971, and used with permission from the Society of American Foresters.)

Airflow from rough to smooth surfaces

When air flows through a forest, the air above the forest canopy moves faster than the air flowing through the canopy, with a strong inflection point near the top of the canopy (Figure 7.12). As the airflow exits the forest and enters an adjacent grassland, there is a mixing zone where the inflection point begins to disappear. It is within this mixing zone where downdrafts occur (Figure 7.13), turbulence increases, and low-level wind currents

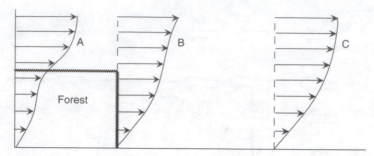

Figure 7.12 Schematic diagram showing the wind profile of airflow (vertical curved line) as it (A) passes through the forest interior, (B) leaves the forest and enters the mixing zone where the airflow adjusts to the smoother surface of the grassland, and (C) enters the zone where the airflow has adjusted to the new surface.

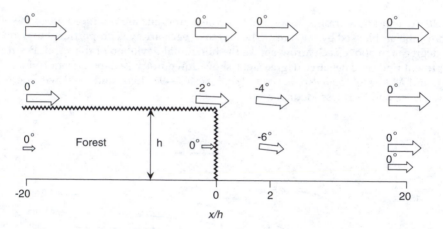

Figure 7.13 As a breeze travels from a forest to a prairie, downdrafts occur at the forest-prairie edge and for a short distance downwind of the edge. The length of each arrow is proportional to wind velocity, and the degree sign above each arrow shows the horizontal deviation of the wind. Positive degree signs represent updrafts; negative degree signs show downdrafts.

that passed through the forest to grassland begin to increase in speed. However, wind currents close to the ground may go for considerable distances ($x > 100h$) before they gain enough speed to adjust fully to the smoother surface of the grassland (Raynor 1971; Gash 1986; Lee 2000).

Impact of turbulence caused by habitat edges on olfactory predators and their prey

Whenever there is a change in canopy height at a habitat edge or ecotone, turbulence will increase, and updrafts will occur if the airflow is moving from the shorter to the taller canopy. For these reasons, olfactory predators will be less efficient when hunting along habitat edges than when hunting within the interior of a habitat patch. This may explain in part why many animals hide along forest edges (Figure 7.14) and why many birds prefer to nest there.

Figure 7.14 Fawns often hide along forest edges. Can you find the second fawn in this photo? (Courtesy of Madsen.)

chapter eight

Turbulence within and below plant canopies

Convective turbulence within plant canopies

The amount of radiant energy exchanged between two surfaces depends on the temperature difference between them. Most of the sunlight is absorbed by the leaves in the forest canopy, resulting in a layer of air in the canopy region that is warmer than the air close to the ground or the air at the height of the tree trunks. Sun and Mahrt (1995) compared temperatures within a black spruce (*Picea mariana*) forest in Canada during midday over a 3-day period of sunny weather in September. They found that the mean temperature in the forest canopy was 27°C; the forest floor was between 19 and 25°C beneath the shade of trees and 25 and 33°C in the sun (Figure 7.11). Hence, there is considerable mixing of air between the sunny and shaded parts of the canopy and between the upper canopy and the boundary layer above it. However, there is less mixing of air between the upper canopy and the subcanopy. In one study of airflow in an old-growth Douglas fir forest, Lee and Black (1993) also found that a temperature inversion occurred on sunny days between the warmer tree canopy and cooler forest floor. This inversion greatly suppressed the vertical movement of air near the forest floor and in the subcanopy.

An exception to this pattern occurs where there are forest clearings or large gaps in the forest canopy (Figure 8.1). Here, sunlight can reach close to the forest floor and warm the air adjacent to the ground. Warm air is drawn upward and replaced by colder air seeping into these clearings from the surrounding forests. Hence, forest clearings act as giant conveyer belts during the day, allowing cool air along the forest floor to be drawn upward.

At night, the movement of air within a forest changes. The leaves in the upper canopy of the forest are exposed to the night sky, which has a temperature of absolute zero (−278°C); other surfaces (for example, leaves, stems, and the ground) will be at ambient temperatures. Hence, leaves in the upper canopy cool faster at night than leaves in the lower canopy. Because of this, the air in the upper canopy becomes colder than the air beneath it. This colder air sinks and is replaced by warmer air rising from the subcanopy space; this process continues throughout the night. In forests where the upper canopy can be several meters above the ground, this nocturnal temperature difference causes a slow but significant flow of air between the cooler air of the upper forest canopy and the warmer air along the forest floor.

Schal (1982) noted that at night the air along the ground of tropical rain forests was 1 to 3°C warmer than air just 2 m above it, and that this phenomenon caused ground-level odorant plumes to rise to heights greater than 2 m. At that height, odorants would

Figure 8.1 On sunny days, forest clearings are warmer than the adjacent forest because sunlight warms the ground. Hence, updrafts are common in forest clearings on sunny days, and these updrafts make clearings good places to hide from olfactory predators. (Courtesy of Russ Kinne and Tony DiNicola.)

be above the olfactory zone of mammalian predators. Interestingly, male cockroaches in tropical forests have to climb up plants at night so that they can detect the rising phero-mone plumes of female cockroaches on the ground (Schal 1982).

Mechanical turbulence within plant canopies

As we discussed, airflow within the boundary layer of the atmosphere is turbulent, with the largest eddies reaching up to the top of the layer. Energy in the large eddies cascades down through even smaller eddies, which have wavelengths of meters and millimeters. It is the larger eddies that cause odor plumes to meander horizontally and undulate vertically. The smaller eddies spread odor plumes apart as they move downwind, and the smallest eddies tear odorant filaments apart, mixing the odorant molecules contained within them with odorant-free air. These different-size eddies are all embedded within each other in a random, chaotic fashion so that it is difficult to predict how a particle or odorant molecule will move once it is released by the odorant source.

Airflow within a plant canopy is different from airflow within the boundary layer. It is still turbulent, but it is not random. The plant canopy is made up of millions of surface features, each of which has a natural wavelength that is twice its length in a stream-wise direction. As air flows through the canopy, it has to bend to go around each of these surface features. In doing so, an eddy develops in the airflow that is the same as the natural wavelength of those surface features. Eddies of other wavelengths are dampened out by the canopy because they encounter more friction than eddies that are of similar wavelength as the surface features (Figure 8.2). The result is that the random, chaotic eddies that occur above the canopy become more ordered within it. That is, turbulent energy is concentrated in fewer wavelengths. A plant canopy is composed mainly of leaves, stems, and twigs with a length often measured in centimeters. Hence, much of the turbulent energy within a canopy will be in same-scale eddies.

Figure 8.2 Beneath forest canopies, thick vegetation and branches dampen out most eddies. (Courtesy of Russ Kine and Tony DiNicola.)

One way to view the turbulence in plant canopies is to view the upper canopy as a region where two columns with air flowing at different velocities come into contact. One of these is the faster-moving air above the canopy, and the other is the slower-moving air within and below the canopy that has lost velocity because of friction with the plant surfaces. The momentum for the air flowing within a plant canopy comes from the airflow in the boundary layer immediately above the canopy. Because of this, canopy airflow is coupled with the airflow above the canopy.

Therefore, mean wind speed U and turbulence (σ_u, σ_v, and σ_w) decline as one moves deeper within the canopy and farther away from the source of the wind's momentum (Raupach et al. 1996). However, U declines faster than turbulence, so that turbulence intensity (σ_v/U, σ_u/U, and σ_w/U) increases (Shaw et al. 1988). The result of this is that the concentration of odorants leaving an animal in a canopy will be more concentrated (number of odorants/cubic meter of air) than if the same animal was in an open prairie because of slower U passing over the odorant source. Odor plumes in canopies will move downwind at much slower velocities than if they were in the open, but as they move downwind, the plumes will disperse more rapidly per meter traveled but not per second because of the higher turbulence intensity in canopies (that is, the angle at which the plume is expanding will be wider). Given that odorants are so light that gravity has little impact on them, measuring turbulence intensity may be more important than measuring turbulence as far as predicting an animal's vulnerability to detection by an olfactory predator.

Although airflow within canopies is entrained with and powered by the airflow above it, airflow within the canopy is much more intermittent than airflow in the boundary layer. Within canopies, there may be periods of several seconds or minutes of little air movement followed by brief, intense incursions of high-velocity air that move downward into the

Forb canopy   Grass canopy

Figure 8.3 Silhouettes of forb and grass canopies showing how the canopy and therefore λ are high within a forb canopy and low within a grass canopy.

canopy from above. Concomitant with these gusts or sweeps, air from the plant canopy is ejected out of the canopy. These *sweeps and ejections* are the dominant wind events within canopies. They are large eddies with a length in a stream-wise direction that is approximately the height of the canopy *h* and with a vertical height that is about *h*/3 (Raupach et al. 1996).

Airflow and turbulence within forb and grass canopies

Wind profiles in forb and grass canopies differ from each other because plants' leaves and stems, including dead ones (and therefore the canopy's frontal area, λ), are concentrated at different heights (Figure 8.3). In forb canopies, most of the λ occurs in the upper canopy ($z > 0.5h$). For instance, in mature soybean fields ($h = 1.0$ m), most λ occurs at a height of $z = 0.5$ to 0.9 m (Baldocchi et al. 1983), and it is within this region where *sweeps* (i.e., downward-directed gusts) lose most of their velocity. Further down inside a soybean canopy where soybean stems are located ($z < 0.5h$), wind velocities do not change with height, and airflow occurs along the ground (Figure 8.4). Wind velocities within a soybean canopy are slower within a plant row than in the space between rows (Baldocchi et al. 1983). This difference results from the soybean plants offering more resistance to the wind within a row where the plants' stiffer stems are located, whereas the stems that extend

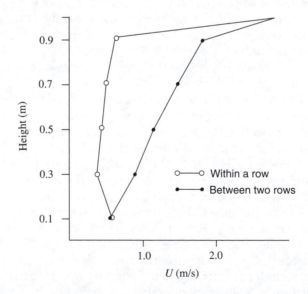

Figure 8.4 Mean wind velocity profiles within a soybean field with a closed canopy of height *h*. Wind velocities within rows are lower than between rows. (Adapted from Baldocchi, D.D., S.B. Verma, and N.J. Roserberg, *Boundary-Layer Meteorol.* 25:43–54, 1983, and used with permission of Springer Science and Business Media.)

out into the space between the rows are thinner, bend easily, and offer less resistance to the wind.

In grass canopies, however, most of the plants' λ are close to the ground. Because of this, a sweep that penetrates into a grass canopy will meet increased resistance as it moves lower into the canopy, and few sweeps will have enough energy actually to reach the ground (Aylor et al. 1993). Another difference between forb and grass canopies is that grass leaves tend to be long and flexible, and a breeze will cause them to bend and orient with the wind. Grass leaves are also angled upward, and once oriented with the wind, they will cause the air flowing within a grass canopy to be deflected upward. This upward deflection of air flowing through grass canopies will cause odorants from an animal hiding in grass to be deflected upward, and its odorants more likely to rise above the olfactory zone of a predator than if the animal was hiding within a forb canopy. Hence, grass canopies should provide safer hiding sites from olfactory predators than forb canopies.

Movement of a pheromone plume within a grain field

Perry and Wall (1986) have studied the movement of pheromones within a wheat field and how male pea moths (*Cydia nigricana*) navigate through a wheat field to find the pheromone source. The authors distributed a series of 3 to 9 insect traps lined up in a stream-wise direction beneath the wheat canopy and baited each with the moth's reproductive hormone. They predicted that the farthest trap downwind should catch the most moths if the moths were following the odor plume back to its source; instead, the farthest trap upwind caught the most moths. Further research showed that the reason for this unexpected result was that, in the dense wheat canopy, discrete odor plumes disintegrate within 10 to 20 m of the odorant source and are replaced by a continuous, homogeneous cloud of pheromones within the wheat field.

The reason for this homogeneous cloud is that the pheromone was constantly absorbed by the surrounding vegetation and then released later (Figure 8.5). For instance, after a pheromone-baited trap was removed, the vegetation surrounding it continued to release sufficient quantities of the pheromone so that males were still attracted to the site for several hours (Perry and Wall 1986). Hence, each of these plant surfaces became a secondary odorant source, and as the wind shifted directions within a wheat canopy, there was a primary pheromone source with a discrete odor plume of a few meters as well as hundreds or thousands of secondary pheromone sources all releasing pheromones at the same time (Figure 8.6). This phenomenon explained why discrete pheromone plumes rapidly give way to homogeneous pheromone clouds within dense wheat canopies.

This phenomenon can also explain why pea moths are caught most often in the upwind pheromone trap. Pea moths fly upwind when they detect the pheromone and continue to do so until they reach clean air that contains undetectable concentrations of the pheromone. Hence, their movements take them to the edge of the homogeneous pheromone plume (Figure 8.7). At that point, they start turning at right angles to the wind (i.e., casting about). In some cases, their casting about causes them to lose the pheromone plume (Moth A in Figure 8.7). In other cases, their casting about causes them to reacquire the plume, and they continue their movements within it (Moths B and C in Figure 8.7). Because moths are moving along the edge of the plume, they are more likely to encounter the most upwind trap rather than a trap located downwind in the middle of the pheromone plume.

This experiment showed that the reproductive pheromone of the pea moth is easily absorbed and then released later by wheat plants, creating a multitude of secondary pheromone sources. Likewise, Karg et al. (1994) showed that the reproductive pheromone for light-brown apple moths (*Epiphyas postvittana*) is absorbed and released by apple

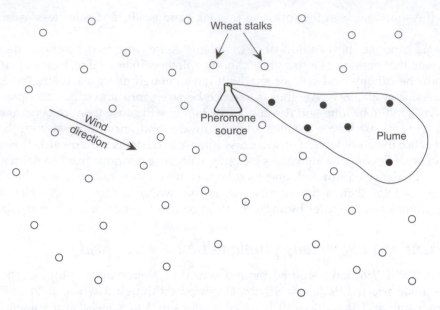

Figure 8.5 A stylized figure of a wheat field showing a single pheromone source, its odor plume, and stalks of wheat that have absorbed some pheromone from the plume (closed circles) and other wheat stalks that have not (open circles).

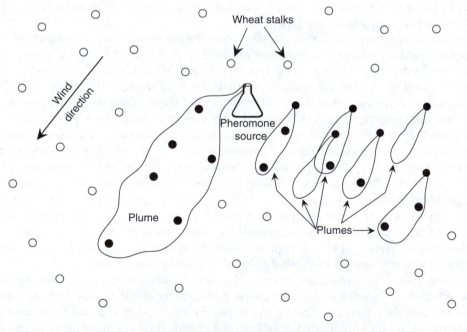

Figure 8.6 The same stylized figure of a wheat field shown in Figure 8.5, but it shows conditions a few minutes later. The wind has shifted now, and all of the stalks that absorbed some pheromone earlier are releasing it and producing their own odor plume.

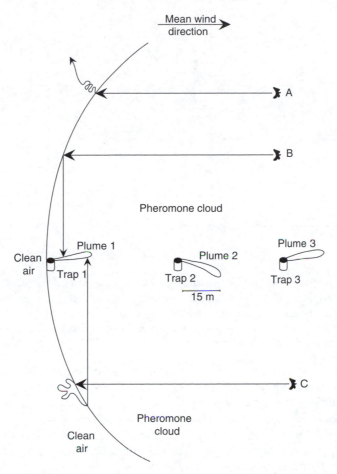

Figure 8.7 Map of a wheat field showing the location of three pheromone traps, the discrete pheromone plume that each is producing, and the homogeneous pheromone cloud. Pea moths A, B, and C all head upwind until they lose contact with the pheromone at the edge of the pheromone cloud. They then start casting about trying to relocate it. The efforts of Moth A are unsuccessful, and it flies off. Moths B and C are more successful, and they reenter the pheromone cloud. Ultimately, Moths A and B discover the discrete pheromone plume of Trap 1 and enter that trap. (Data from Perry, J.N. and C. Wall, in T.L. Payne, M.C. Birch, and C.E.J. Kennedy, Eds., *Mechanisms in Insect Olfaction*, Clarendon Press, Oxford, U.K., 1986, pp. 91–96.)

leaves. The extent to which the odorants used by predators to locate prey are absorbed and released again by vegetation is unclear, but it probably does happen to some extent. Hence, an animal hiding in dense vegetation that exceeds its height probably produces an odor plume that has less-discrete edges and is more homogeneous than if it were in the open because of the absorption and later release of its odorants by surrounding vegetation.

Airflow within the subcanopy of forests

Forest canopies differ from grass and forb canopies in that they are often located many meters above the ground. Hence, z_0 is usually located somewhere in the forest canopy high off the ground, and d, the distance between the ground and z_0, is large. In oak forests, 70% of leaves are in the upper 20% of the canopy (above $0.8h$), and it is in this upper 20%

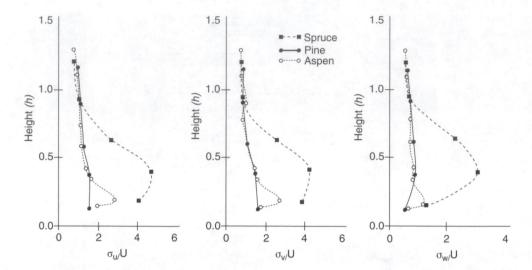

Figure 8.8 How turbulence intensities increase beneath the canopies of a black spruce, jack pine (*Pinus banksiana*), and aspen (*Populus tremuloides*) forest. The leaf area index for the three forests was 10, 4, and 2, respectively, and this accounts for the higher turbulence intensities in the spruce forest because its thick canopy resulted in a lower U. (Adapted from Amiro, B.D., *Boundary-Layer Meteorol.* 51:99–121, 1990, and used with permission of Springer Science and Business Media.)

of the canopy where 90% of the wind's momentum is absorbed by the leaves and branches (Baldocchi and Meyers 1988).

Wind velocities in the subcanopy space of forests are only a fraction of those that occur above the canopy. Mean wind velocities U in the subcanopy space of forests without an understory are 50 to 80% lower than simultaneous velocities above the canopy compared to reductions of more than 90% in forests with dense understories (Amiro 1990; Lee and Black 1993). Turbulence intensity is also higher within the subcanopy space than above the forest canopy (Figure 8.8) or over open ground (McBean 1968; Baldocchi and Meyers 1988; Amiro 1990). Airflow in the subcanopy of a forest is dominated by the same type of eddies (i.e., sweeps and ejections) that occur in forest canopies.

Airflow within the subcanopy space of forests that lack dense understory vegetation differs from airflow within forests that have it. When there is a lack of understory vegetation, the forest is often described as park-like, and a person looking horizontally sees mainly the trunks of the canopy trees (Figure 8.9). These conditions often occur when the canopy of the dominant trees is closed so that insufficient light reaches the forest floor to support understory vegetation. Henceforth, I refer to these park-like forests that lack an understory as *closed-canopy forests*. In such forests, the vertical profile of wind velocities is bimodal. Velocities are highest above the forest canopy, with a secondary peak in velocity in the open space beneath the forest canopy where only tree trunks occur (Figure 8.10). This secondary peak does not occur in forests where there is dense vegetation in the subcanopy space and along the ground.

In closed-canopy forests, the surface features that influence airflow consist of large tree trunks that may be spaced meters apart (Figure 8.10) rather than by centimeters that separate the surface features of grass habitats. In such forests, there is often a bimodal size distribution of eddy sizes in the subcanopy. The larger eddies are associated with the sweeps, and their sizes are influenced by h and turbulent events occurring above the subcanopy space. They rotate around a horizontal axis and produce vertical turbulence. The smaller eddies are associated with the wakes created by airflow around the tree trunks.

Figure 8.9 In open forest, the understory is dominated by large tree trunks, and understory vegetation is sparse. (Courtesy of Madsen.)

These eddies rotate around a vertical axis and cause stream-wise and lateral turbulence.
The sizes of these eddies are approximately twice the diameter of the tree trunks (Lee and
Black 1993).

   Atmospheric stability influences turbulence within forest subcanopies. In a deciduous
forest, vertical turbulence σ_w is lower when the atmosphere is stable than when it is either
neutral or unstable (Leclerc et al. 1991). During stable conditions, sweeps rarely penetrate
far into the subcanopy space, and there is little vertical turbulence close to the forest floor.
Lateral and stream-wise turbulence are also suppressed near the forest floor under stable
atmospheric conditions.

Differences in the movement of odor plumes above grass canopies and within forest canopies

David et al. (1982, 1983) studied the flow of soap bubbles released over an open field and
found that, once released, most bubbles floated in a linear fashion for many meters. The
path connecting a string of bubbles released a few seconds apart often was bent and
twisted, but this was mostly because of a change in wind direction at the odorant source,
so that bubbles released later traveled on a different trajectory than earlier bubbles (Figure
8.11A). In such areas, a male moth can find a pheromone source by simply flying upwind

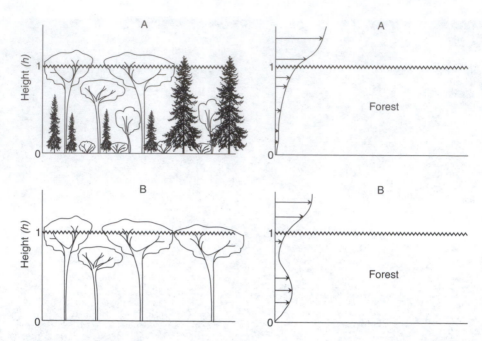

Figure 8.10 Wind profiles of (A) an open-canopy forest with an extensive forest understory and (B) a closed-canopy forest that has an open subcanopy.

when it detects a pulse of reproductive pheromones in a breeze (David et al. 1982, 1983). The same is also true for predators trying to locate an animal by its odor plume.

Within forests, soap bubbles behave differently from those released over an open field (Figure 8.11B). Rather than floating in straight lines for several meters, soap bubbles released in a forest changed their trajectory by more than 90° before they had traveled more than 20 m from the release site (Elkinton et al. 1987). Moreover, the bubbles in the plume all changed direction concurrently.

This difference in the behavior of bubbles above open fields and within forests is easily explained. Bubbles released above an open field move at the same speed as the wind in the boundary layer, and they become entrained in that gust or parcel of air that existed at the movement of their release. They continue to travel in the same direction and speed as that gust or air parcel. In contrast, airflow beneath a forest derives its energy and direction from the air flowing above the canopy, but because of friction, its velocity is much slower than the air flowing above the forest canopy. The gust of air above a forest canopy that exists when a bubble is released beneath the canopy provides the initial direction and velocity for that bubble. However, that gust quickly passes by and leaves the slower-moving bubble behind. Because the bubble's energy is no longer coming from the initial gust, that bubble and other nearby bubbles change direction and speed each time a new gust passes overhead.

Odorant and pheromone filaments will behave similarly to soap bubbles, and their constant changing of directions within a forest canopy means that if male moths located in a forest fly upwind whenever they detect a pheromone, they will often be flying away from the odorant source (Carde 1986). This also means that an olfactory predator will be unlikely to locate a quarry hiding in a forest by simply moving upwind.

However, there are methods that olfactory predators located in forests can use to increase the probability that when they head upwind on detection of an odor plume, they are actually heading toward the odorant source. On days when the wind above the canopy is slow ($U < 9$ km/h), the air within a forest canopy and subcanopy will stop moving for

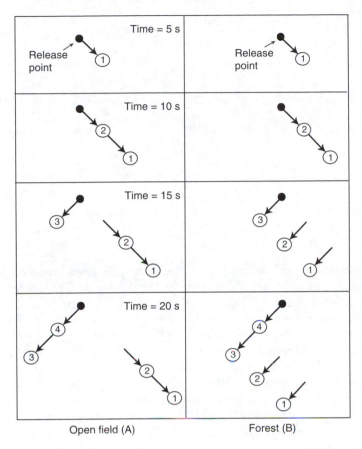

Figure 8.11 Movements of four soap bubbles during consecutive 5-second intervals. Each bubble is individually numbered, and bubbles were released every 5 seconds from a source either above an open field or beneath a forest canopy. In an open field (A), soap bubbles generally followed the same direction through the entire period and traveled in different directions from each other. In contrast, soap bubbles in a forest (B) generally all moved in the same direction at the same time.

many seconds or minutes. These quiescent periods are often followed by a light breeze that lasts a few seconds. These short-duration breezes often deviate from the direction of the wind above the canopy and shift from one direction to another.

At other times within a forest canopy, breezes persist for many seconds, blow in a constant direction, and are aligned with the boundary layer wind above the plant canopy (Elkinton et al. 1987). A possible explanation for these different wind patterns comes from the sweeps and ejections that dominate within plant canopies. When sweeps occur, they will be aligned with the direction of airflow above the canopy. A sweep has a curved frontal edge and displaces the canopy and subcanopy air in front of it. Air at the two sides of the sweep is displaced away from it at a 45 to 90° angle to the stream-wise direction that the sweep itself is moving (Figure 8.12). This displacement of wind from the sides of a sweep accounts for the short-duration breezes observed by Elkinton et al. (1987). However, most of the sweep's energy will be at its center, and it may have enough energy to cause stream-wise airflow for a considerable distance downwind from where it entered the canopy. The long-duration breezes noted by Elkinton et al. (1987) may have occurred when they were directly downwind from where a sweep entered the canopy.

Elkinton et al. (1987) hypothesized that male gypsy moths located within a forest would rarely find the female if they flew upwind during one of the short-duration breezes

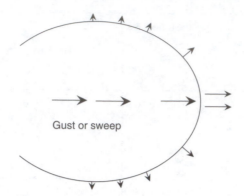

Figure 8.12 As a gust sweeps down into a forest canopy, it pushes forward a small amount of air away from its sides and a larger amount of air from its front. Because sweeps are curved surfaces, a single sweep will cause air beneath a forest canopy to move in different directions.

because they would usually be heading in the wrong direction. To avoid this problem, moths should only take wing and fly upwind during a long-duration breeze. This strategy should increase the probability that their flight will take them toward the pheromone source. Even so, few male moths located within forests were able to locate females by following pheromone plumes if they were more than 80 m away from the female, although they could detect the pheromone at much greater distances (Elkinton et al. 1987).

The meandering of soap bubbles or odor plumes within forests decreases as wind velocities increase. Brady et al. (1989) demonstrated this by plotting the path of smoke puffs or soap bubbles every 2 seconds over an open field and in a forest. They found that as wind velocity increased, the change in direction between consecutive paths decreased (i.e., their paths became straighter). This was especially pronounced within forests (Figure 8.13).

How does turbulence within a forest plantation differ from a naturally reproducing or old-growth forest?

Forest plantations differ from naturally reproducing or old-growth forests because their trees are of the same age and similar height. Because of this, the top of the forest canopy is smooth. There are also few gaps caused by tree falls or clearings. Because the top of the forest canopy is so even, these forests have a small z_0, and the wind flows smoothly above the forest canopy. Hence, within the forest canopy and subcanopy space, there are few sweeps, little or no wind, and little turbulence.

The forest canopy of naturally reproducing or old-growth forests is composed of trees of many ages and heights. For this reason, these forests have a larger z_0 than forest plantations. Gaps in the forest are numerous where the old trees have died or fallen. Airflow at the upper parts of an old-growth forest is turbulent, and gusts are often directed downward. One reason for this is because trees that rise above the rest of the canopy act as isolated surface features. As air flows around these tall trees, their resistance to airflow directs some of the airflow downward into the forest canopy.

Gaps in the forest canopy and clearings cause air moving above the forest canopy to drop down into the clearing and through the forest edge downwind of the gap. This increases wind velocities both in the forest gap and in the downwind forest. Wind patterns within a forest clearing often exhibit U-shaped streamlines and eddies (Figure 8.14), with the latter more likely to occur when ambient winds increase in velocity (Bergen 1975;

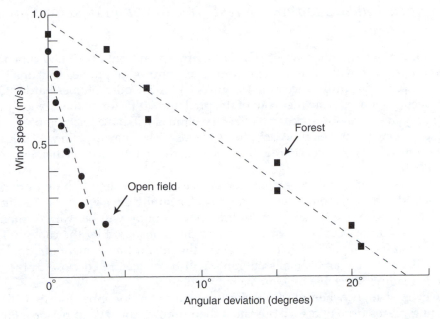

Figure 8.13 Relationship between wind velocity and the angle between two sequential paths taken by a soap bubble when each path was 2 seconds in duration. An angle of 0° means that the bubble maintained a straight line across the two paths. (Adapted from Brady, J., G. Gibson, and M.J. Packer, *Physiol. Entomol.* 14:369–380, 1989, and used with permission of the Royal Entomological Society.)

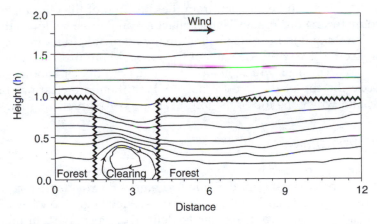

Figure 8.14 Side view showing airflow streamlines in a clearing surrounded by an old-growth spruce-fir forest. The dark horizontal lines shows where the forest is located; its absence marks the clearing. (Adapted from Miller, D.R., J.D. Lin, and Z.N. Lu, *Agric. Forest Meteorol.* 56:209–225, 1991, and used with permission of Elsevier.)

Lee 2000). Eddies are also more likely where gaps are small and surrounded by tall forests (that is, when gaps are box-like in appearance).

Stacey et al. (1994) observed that the probability of gusty winds occurring in a forest interior is doubled when the length of a canopy gap is greater than $2h$. These gusts are often sufficient to topple other trees, causing small gaps to grow larger. In addition to being a site where mechanical turbulence directs gusts downward, forest clearings allow sunlight to penetrate to the forest floor, and because of this, they are sites of updrafts on sunny days.

Impact of turbulence within a forest subcanopy on olfactory predators and their prey

The intermittent gusts, which sweep the subcanopy space from above and concomitantly eject subcanopy air, will have a major impact on any odor plumes that may exist in subcanopy space. The turbulence from the gusts will make odor plumes harder to follow, and the ejections may elevate them out of the predator's olfactory zone. Hence, predators tracking prey in subcanopy spaces via olfactory plumes will have to accomplish their task before the next sweep and ejection complicate the task. This may explain why dogs trained to find human cadavers have a harder time locating them in the woods than in open fields (Komar 1999).

Forests, however, can also be dangerous places for an animal to hide from olfactory predators under certain atmospheric conditions. On sunny days when there is little wind or when the atmosphere is stable, air in the upper canopy will be warmer than the air along the forest floor because most of the sunlight is absorbed by the canopy. This will cause a temperature inversion that will prevent the upward movement of odorants from the forest floor; rather, an animal's odorants will be concentrated close to the ground, where it is readily detectable by mammalian predators.

However, at night when most olfactory predators are foraging, forests will provide safer hiding places from olfactory predators than prairies for several reasons. First, horizontal wind velocities in forests, especially those with a subcanopy, are more likely to be at suboptimal levels for olfactory predators than open fields. Second, air along the floor of forests is often warmer at night than air higher up in the subcanopy space, and this causes odorants along the ground to rise to above 2 m, well above the olfactory zone of most mammals. Third, the continually shifting wind direction in forests means that it is unlikely that the wind will be blowing from the direction of the odorant source when a predator encounters an odor plume. Hence, olfactory predators in a forest that are located many meters from an odorant source will normally be unable to locate the source.

Prey will also be safer in those parts of the forest where these sweeps and ejections are most likely to occur. This will be in areas where a high tree or another surface feature protrudes above the top of the forest canopy, where there are clearings or gaps in the forest canopy, and where the top of the forest canopy is rough or irregular (z_0 is high).

Airflow in savannas

Savannas are normally defined by ecologists as areas with an extensive grass or forb cover and scattered trees or brush. By this definition, savannas are diverse; hence, they are often subdivided into categories based on the amount of wooded vegetation they contain (Figure 8.15). In terms of turbulence, savannas can be defined as open, bare, or grassy areas that also contain isolated surface features (trees or bushes). The trees or bushes protrude above the grass canopy high enough to create turbulence and are dispersed far enough apart that the effect of these individual trees on airflow is discernable. How high or far apart surface features must be to satisfy this requirement has not been determined. As a rule of thumb, I propose that to qualify as a savanna the height of surface features must be 2 m high and at least five times as high as the ground cover, and that the mean distances between the trees must exceed $2h$.

One important question to consider is what distance between surface features maximizes turbulence and wind velocity along the ground of savannas. A seminal study on this topic was conducted by Green et al. (1995), who examined airflow and turbulence in a series of plots within a Sitka spruce (*Picea sitchensis*) forest where trees on average had a canopy of 2 m in diameter and a height of 8 m. Hence, the trees were tall and thin as

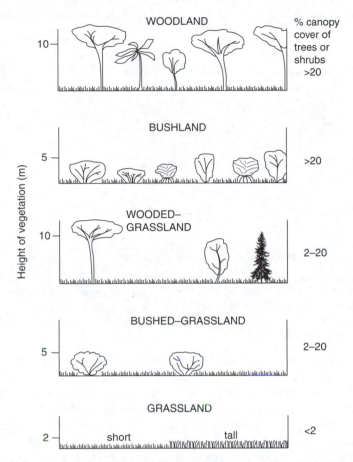

Figure 8.15 Classification of savannas into different habitat types based on the amount of woody vegetation they contain. (Adapted from Deshmukh, I., *Ecology and Tropical Biology,* Blackwell Scientific Publications, Palo Alto, CA, 1986, and used with permission of the Royal Entomological Society.)

this is the general characteristic of Sitka spruce grown under forest conditions. In the plots, the trees were thinned so that the few remaining trees were spaced in rows with a gap of 4, 6, or 8 m between both rows and trees. The thinning reduced the tree density from 3000/ha to a density of 625, 278, or 156 trees/ha. Individual trees in rows of 4, 6, and 8 m had mean frontal area indices λ of 1.0, 0.5, and 0.25, respectively.

Green et al. (1995) found that the wind speed in the trunk space was reduced 54% in the 8 × 8 m spacing, 71% in the 6 × 6 m spacing, and 84% in the 4 × 4 m spacing as compared to wind velocity values above the canopy. Turbulence (σ_u, σ_v, σ_w) below the canopy increased with decreasing distance among trees (Figure 8.16). The increase in tree spacing allowed airflow momentum to penetrate lower into the canopy and thereby enhanced turbulent exchange close to the ground. The penetration depth, which Green et al. (1995) defined as the height where 50% of the momentum had been absorbed by the canopy, was $0.75h$ at the 4 × 4 m spacing, $0.70h$ at the 6 × 6 m spacing, and $0.61h$ at the 8 × 8 m spacing. In contrast, this depth was $0.86h$ in a deciduous forest (Shaw et al. 1988). These findings demonstrate that there is greater penetration of wind into the canopy spaces of savannas where trees are widely spaced apart than into the canopy of forests.

As in forests, maximum wind velocities and turbulence occur in the region just above the canopy in savannas. In savannas, wind velocity and turbulence both decline with height above the ground but do so at a slower rate than in forests, which is not surprising

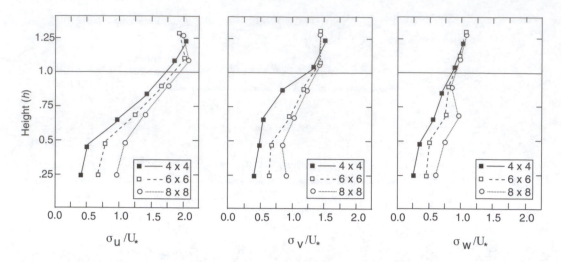

Figure 8.16 Impact of tree density in a pine plantation on turbulence intensity in a stream-wise direction σ_u/U, horizontal turbulence intensity σ_v/U, and vertical turbulence intensity σ_w/U when measured at different heights within the subcanopy. A height of 1.0 marks the top of the canopy. (Adapted from Green, S.R., J. Grace, and N.J. Hutchings, *Agric. Forest Meteorol.* 74:205–225, 1995, and used with permission of Elsevier.)

given the more open nature of savannas. Savanna winds are often channeled through gaps among the trees (Figure 8.17). It is in such gaps rather than in the wakes behind trees where stream-wise velocity U and turbulence are highest, but wind velocity in these gaps depends on the gap's length, width, and orientation relative to U. Trees in savannas also cause convective turbulence because of differential temperatures between the shaded ground beneath the trees and the open ground beyond their canopy. Such temperature differences can occur either during sunny days or clear nights and result in small-scale circulation patterns (Lee 2000).

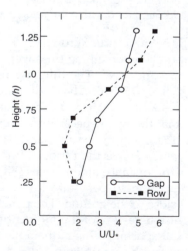

Figure 8.17 Vertical distribution of wind velocity U in a spruce plantation when measured in a gap between rows and within a row of closely spaced trees. (Adapted from Green, S.R., J. Grace, and N.J. Hutchings, *Agric. Forest Meteorol.* 74:205–225, 1995, and used with permission of Elsevier.) The trees were 8 m tall and skinny because they had recently been thinned from 3000 trees/ha to 156 trees/ha. Wind velocity has been normalized by the friction velocity (U_*), and $h = 1$ is the top of the canopy.

Sidebar 8.1: Why Did African Wild Dogs Become Extinct in the Serengeti Plains But Not Elsewhere in Africa?

In the open plains of the Serengeti, packs of wild dogs (*Lycaon pictus*; Figure 8.18) hunt large herbivores weighing more than 10 kg. The wild dog population declined from over 100 individuals in the 1960s to about 40 in the 1970s and became extirpated by the 1990s (Carbone et al. 2005). During this same period, spotted hyena populations on the Serengeti increased, and the smaller wild dogs were unable to protect their kills from the larger hyenas. Many researchers believed higher hyena densities meant that wild dogs had less time to consume a kill before a hyena found the carcass and displaced the wild dogs (Fanshawe and Fitzgibbon 1993; Carbone et al. 1997, 2005). Yet, wild dog populations are doing well in other parts of Africa where forests and wooded savannas occur despite the presence of hyenas (Creel and Creel 2002).

Why are wild dogs more vulnerable to hyena stealing their kills in open plains? The answer is that, within open plains, hyenas can quickly locate a carcass by watching vultures (Kruuk 1972) or by following the odor plume to a carcass (Figure 8.19) (Mills 1990). Within wooded areas, it takes longer for hyenas to locate a carcass because vultures cannot easily spot carcasses through the trees, and hyenas have more difficulty following odor plumes in wooded savannas. As noted in this chapter, airflow in savannas will be more turbulent than in open plains. Hence, wild dogs have more time to feed at kill sites in wooded savannas than in open plains. The extra time at their kill sites seems to make the difference in areas where they can survive despite hyenas.

Figure 8.18 An African wild dog. (Courtesy of U.S. Department of Agriculture's Wildlife Services.)

Figure 8.19 In open habitat, hyenas can quickly locate kills made by wild dogs by use of their great olfactory acuity and by watching vultures. (Courtesy of Johan du Toit.)

Impact of turbulence in forests, prairies, and savannas on olfactory predators and their prey

In prairies, animals will be safer from olfactory predators if they hide in areas that have a rough surface (high z_0), such as fields with tall forbs and grasses, hilly terrain, or boulder-strewn areas. Animals should also hide where nocturnal wind is either above or below optimal speeds for olfactory predators and where updrafts can be expected, such as on south-facing slopes.

When animals are hiding from olfactory predators in forests, they should be safer in even-aged forests that have thinned rather than unthinned stands, in old-growth forests rather than in even-aged forests, in forests where the upper edge of the forest canopy is irregular in height, or downwind of forest clearings. All of these locations will cause airflows to be directed downward into the subcanopy.

The scattered trees and bushes in savannas can act as isolated surface features and help an animal's odorant trail to disperse more rapidly than on an open plain. Yet, if savannas become too thick with trees and brush, their canopies will create a displaced surface, and wind will not penetrate to the ground. Savannas provide good places for animals to hide from olfactory predators when trees and bushes are scattered so far apart that they act as isolated surface features. Hence, it is not surprising that greater sage-grouse (*Centrocercus urophasianus*) prefer to nest in patches of sagebrush (*Artemisia* spp.) with an intermediate level of brush (15 to 25% canopy cover) and avoid nesting in stands where the sagebrush is either too sparse or too thick (Connelly et al. 2000; Holloran et al. 2005).

In savannas, the best hiding places for animals are in the gaps or spaces between the scattered trees and bushes, especially gaps that are aligned with the wind direction. In such areas, wind velocity near the ground will be highest.

chapter nine

Trade-offs required to achieve optimal hiding strategies

Thus far, I have made predictions about how animals should respond so that they can reduce their probability of being detected and located by an olfactory predator. In reality, the choices an animal must make to avoid predation are much more complicated than I have suggested because visual predators also threaten animals, and survival is influenced by more factors than those related to avoiding detection. In this chapter, I describe some of these other factors that influence an animal's chance of survival and the trade-offs that result from them.

Optimal hiding strategies for prey

To be successful, an animal must be able to survive long enough to reproduce, and this means that it must avoid being killed by a predator. If it cannot, then its genes will not be passed along to the next generation. To survive, an animal must have the physical attributes that allow it to avoid predation (for example, camouflage coloration, speed to outrun its predators, weapons to defend itself, and so forth). It must also make correct decisions to avoid predation, and one of these decisions is where best to hide so that its presence cannot be detected by predators. The suite of decisions or rules that animals should follow to minimize their risk of predation is called the *optimal survival strategy*.

Living individuals are the descendents of those individuals from prior generations that made correct decisions about how to avoid predation. Hence, most individuals in a population should be following the optimal survival strategy that was successful for that particular species in the past.

Optimal foraging strategies for predators

Evolutionary forces also shape predators and their foraging decisions. That is, a predator must be able to overcome the defensive mechanisms and the optimal hiding strategy of its prey, or it will starve to death. Evolution is simultaneously shaping prey to increase their ability to avoid predation and shaping predators to overcome the defensive abilities and strategies of prey. Hence, predator and prey are locked in an evolutionary struggle. This has caused coevolution in predator and prey by which the latter develop an effective defensive mechanism and predators develop an effective mean to defeat it.

In this evolutionary struggle, however, predators have an advantage over prey in that they can specialize more easily. For instance, olfactory predators need only refine their unique abilities to locate prey using this modality, and visual predators can concentrate

Figure 9.1 Evolution has provided predators, such as this bobcat, with excellent sensory ability that it can use to find prey, such as this lamb that it just killed. (Courtesy of Madsen.)

on improving their visual acuity. In contrast, prey have to develop the ability to detect not only olfactory predators, but also auditory predators (for example, owls) and visual predators. Hence, prey need to be generalists in their sensory abilities (that is, they need to have good senses of smell, hearing, and vision), but in doing so, their senses will be less finely honed than those of the predators that seek them (Figure 9.1).

How predators develop search images of prey

When a person searches a mountainside for a particular item, such as a deer, it may take considerable time to locate. However, after the first one is discovered, the person learns the deer's color, size, and shape. The person then searches the mountain for objects of similar appearance and in similar areas and through this process is able to find subsequent deer much faster than the time required to find the first one. Thus, the person has developed a visual search image of deer. Based on past hunting experiences, predators also develop a visual search image (Soane and Clark 1973; Pietrewicz and Kamil 1979; Gendron 1986; Nams 1997) and an olfactory search image of prey (Figure 9.2).

Predators increase their probability of finding prey by identifying areas where prey densities are high and by spending more time hunting in these areas. Once there, they increase the thoroughness of their search and decrease their speed (Gendron and Staddon 1983; Guilford and Dawkins 1987; Ibarzabal and Desrochers 2004). Once they have located a prey animal, they restrict their search to the surrounding area because sometimes animals are clumped together rather than distributed in a random or regular pattern. For instance, a clumped distribution would occur if optimal hiding habitat is limited and many animals are trying to hide in the same habitat patch.

Predators can also recall sites where prey were previously located and can increase their hunting efficiency by revisiting these sites during future hunting trips. As an example, European pine martens (*Martes martes*) are able to remember where birds nested from one year to the next and recheck past nesting sites during each breeding season. For this reason, pine martens are more likely to depredate the nests of black woodpeckers (*Cryocpus martius*) that reuse an existing cavity rather than excavating a new one (Nilsson et al. 1991). Tengmalm's owl (*Aegolius funereus*) often nests in nest boxes; predation rates on their nests by European pine martens increase with the age of nest boxes (Sonerud 1985). In an elegant experiment, Sonerud (1989) relocated old nest boxes and found that it was

Figure 9.2 Predators, such as this hunting dog, develop both a visual and an olfactory search image of their prey, in this case a ring-necked pheasant.

the location itself that the pine martens remembered and not some characteristic of older nest boxes that made them easier for pine martens to locate (Figure 9.3).

How birds learn where to nest

Many avian species are more successful nesting in one habitat rather than in another because their morphological and behavioral traits are more suited for particular habitats. Such birds often have a mental image of what an ideal nest site looks like and seek places that best match that image. Yet, birds do not have to wait for evolutionary changes to occur before they adapt to the foraging strategies of predators.

Birds learn from their prior nesting experiences, accumulate knowledge about predators and their hunting methods, and then change their behavior accordingly. For example, birds nesting for the first time are less selective in choosing nest sites than are experienced nesters (Klopfer and Ganzhorn 1985; Thogmartin 1999). Unsuccessful nesters are also less likely to return to their former nest sites the next year than are successful birds (Harvey et al. 1979; Greenwood 1980; Herlugson 1981; Hass 1998).

Nesting birds can also acquire information about predators by watching interactions between their conspecifics and predators (Conover 1987). For instance, ring-billed gulls (*Larus delewarensis*) and California gulls (*Larus californicus*) will abandon former colonies or those parts of the colony where incubating gulls were depredated the prior year (Figure 9.4). They will also abandon colonies when even a single predator is present at the colony site during the prelaying period (Conover and Miller 1978). For solitary-nesting birds, the task of avoiding unsafe nesting areas is more difficult because it is harder for them to observe other nesting birds as they interact with predators. This task is also difficult for birds when predators are located in every potential nesting habitat, and the probability of nest predation increases with predator abundance and activity (Sidebar 9.1). In such situations, birds have to assess predator densities across areas, but it is uncertain how well birds can do this (Moller 1988; Sieving and Willson 1998).

Figure 9.3 European pine martins remember the location of nest boxes used by nesting birds from one year to the next. (Courtesy of Madsen.)

Figure 9.4 Ring-billed and California gulls will abandon colonies once mammalian predators reach them. (Courtesy of Don Miller.)

Sidebar 9.1: Patterns of Mammal Abundance and Their Impact on Nesting Success of Passerines

Cain et al. (2006) wanted to determine if there were safe places for passerines to nest in the mountain meadows of California by assessing where mammalian predators were abundant and active. They found that no place was safe because predators were everywhere. Short-tailed weasels were most common near willows; Douglas squirrels (*Tamiasciurus douglasii*), golden-mantled ground squirrels (*Spermophilus lateralis*), and chipmunks (*Tamias senex, Tamias speciousus, Tamias amoenus,* and *Tamias quadrimaculatus*) were most common along the meadow's edges. Mice (*Peromyscus maniculatus, Reithrododontomys megalotis,* and *Microtus* spp.) and long-tailed weasels (*Mustela frenata*) were active throughout the meadow. Other mammalian predators that commonly hunted in the meadows included mink (*Mustela vison*), American pine marten (*Martes americana*), striped skunk, raccoon, and yellow-bellied marmots (*Marmota flaviventris*). Given this diversity of mammalian predators, it is not surprising that over half of the passerine nests were depredated. Yet, there were different patterns of predation among passerines based primarily on where each species nested in the meadow and which predator was most active in that area. Thus, predation rates on the nests of dusky flycatchers (*Empidonax oberholseri*) and willow flycatchers (*Empidonax traillii*) increased with increasing activity of short-tailed weasels, and nest predation rates on yellow warblers (*Dendroica petechia*) were correlated with the numbers of chipmunks and short-tailed weasels.

It is unclear if a bird that lost one nest to a predator can identify which nest site characteristics contributed to the failure and avoid those same characteristics when it renests. For instance, would a duck that lost its nest because the nest was located on top of a fox den realize that this was the mistake and avoid fox dens in the future, or would it key in on another nest characteristic, such as distance to a forest edge? There is evidence that birds can in some cases identify why a nest was unsuccessful and take steps to reduce that risk when renesting immediately following a failed nesting attempt. For example, pinyon jays (*Gymnorhinus cyanocephalus*) alter the location of the renesting attempt based on the reason why their first nest failed. If a nest failed because of cat predation, then new nests were built higher in the canopy (Marzluff 1988). If a nest failed because of predation by ravens or American crows, then the new nests were built lower in the canopy and in more concealed locations. However, a year later jays returned to building nests high in the canopy because of poor memories, because nest height is a poor predictor of nest success, or because they are more concerned about mammalian than avian predators (Caro 2005).

Another technique that birds can use to identify safe nesting sites is to travel widely at the beginning of each nesting season in search of optimal nesting habitat. Birds that do so are more likely to nest successfully than those that do not (Badyaev et al. 1996). As they travel about, they can watch for predators. Once they have identified a potential nest site, they can spend time there and can make themselves conspicuous while at the potential nest site. If their activity attracts predators, then they can abandon that site and keep searching until they find one that seems to be devoid of predators (Nolan 1978).

Interplay between a predator's optimal foraging strategy and a prey's optimal hiding strategy

Predators constantly refine their decisions about where to hunt based on where their prey are hiding, and prey constantly refine their decisions about where to hide based on where predators are hunting. In this regard, predators have a major advantage. Prey have to make their decision about where to hide first, but predators get to make their decision of where to hunt last. Hence, predators can change their search image of where prey are located faster than the prey can change their image of where it is safe to hide. This disadvantage for the prey is particularly keen for animals that are relatively immobile once they select a hiding place. For instance, upland-nesting birds cannot move their nest once egg laying begins (Figure 9.5). Because eggs require 3 to 4 weeks to hatch once incubation begins, predators will have several weeks to search for them. If all nesting birds one year selected the same nesting strategy (for example, nesting on hilltops), then predators would quickly learn this and limit their searches to such areas. Hence, an area where it was safe to nest one year may be unsafe the next year.

Given this circumstance, what is the optimal hiding strategy for prey? One answer is to avoid nesting close to other nesting birds because nest predation rates increase when nests are clumped together or when two nests are close to each other (Caro 2005). Another answer is to hide where one has been successful in the past in avoiding detection, but once found, to seek a new hiding place randomly and to stay there until discovered. For nesting birds, this would mean returning to the same site used last year, providing that the nesting attempt was successful. Unsuccessful birds or birds nesting for the first time should select nest sites at random from among those sites that appear appropriate based on their earlier experiences or perceptions formed during their early days of life (Klopfer and Ganzhorn 1985).

Figure 9.5 Ground-nesting birds, such as this American avocet (*Recurvirostra americana*), have a difficult task of selecting a safe nesting site in areas where there are high densities of predators. (Courtesy of the U.S. Fish and Wildlife Service's Bear River Migratory Bird Refuge.)

Figure 9.6 Forster's terns (*Sterna forsteri*) prefer to nest in open areas where they have a clear view of their surroundings. (Courtesy of U.S. Fish and Wildlife Service's Bear River Migratory Bird Refuge.)

Trade-offs involving avoiding detection versus capture

For many species, the primary defense against predators is to avoid detection. However, once detected, animals can still survive if they avoid capture. For instance, thick vegetation will simultaneously conceal an animal and impede its ability to keep an approaching predator under observation. Hence, optimal cover may represent a trade-off between these competing needs (Figure 9.6). For example, song thrushes (*Turdus philomelos*) and evening grosbeaks (*Coccothraustes vespertinus*) select intermediate levels of cover for their nest sites. Bekoff et al. (1989) and Gotmark et al. (1995) believe that these birds did so because of the trade-off between the birds' dual needs for concealment and a view of their surroundings.

Thick cover can also impede the escape attempts of incubating birds, and this can influence nest selection behavior. White-tailed ptarmigan (*Lagopus leucura*) nests located in thick cover were less likely to be depredated than those placed elsewhere, but the hens incubating these nests were more likely to be killed because they were less able to escape once detected (Wiebe and Martin 1998). Hence, the benefit of nesting in thick cover is increasing the survival rate of this year's offspring, but this benefit is offset by the higher risk to the hen and thus the loss of her potential to produce offspring in future years.

Trade-offs required to avoid both visual and olfactory predators

To survive, animals must avoid detection by both visual and olfactory predators (Figure 9.7). It is often hard to do both because the optimal hiding places to avoid visual and olfactory predators are not always the same. As an example, open fields with scattered

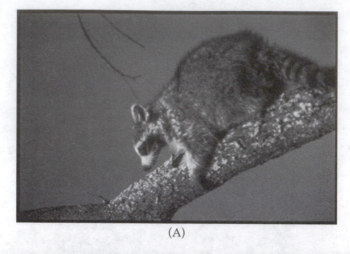

(A)

(B)

Figure 9.7 Animals must avoid both olfactory predators such as raccoons (A) and visual predators such as common ravens (B). (Courtesy of U.S. Department of Agriculture's Wildlife Services.)

trees are better places to hide from olfactory predators than are open fields without trees because the trees will produce turbulence and updrafts. Yet, many visual predators, such as raptors and crows, hunt from perches where they can scan their surroundings, and isolated trees provide them the necessary perches. Hence, the risk of predation from visual predators will be higher in open fields with scattered trees than in open fields that lack them (Preston 1957; Paton 1994; Wolff et al. 1999). Finding a safe nesting site can be particularly problematic in areas where there is a diverse array of visual and olfactory predators, especially if the relative abundance of predators varies across time. In such areas, there may be no nest site characteristic or habitat patch characteristic that birds can reliably utilize to identify where their nests would be safe from predators (Filliater et al. 1994; Jimenez 1999; Rodewald and Yahner 2001).

In areas where both visual and olfactory predators are a threat, the optimal hiding strategy might be to use intermediate sites that offer at least some protection from both types of predators. For instance, animals could hide in areas that contain bushes rather than trees because these areas would still create turbulence and offer protection from olfactory predators. If the bushes are low, then they will provide only poor perch sites for raptors.

Every species varies in its vulnerability to visual and olfactory predators. Nests of small birds may be particularly vulnerable to crows, jays, and magpies; the nests of large birds are less threatened by crows and jays because large birds can more successfully defend their nests from these birds. In contrast, small birds may be harder for olfactory predators to locate because their small size reduces the amount of odorants they emit; larger birds produce more odorants and are a more rewarding target (that is, a bigger meal) for olfactory predators. For instance, midsize mammals (for example, raccoons, opossums [*Didelphis virginiana*], skunks, and foxes) rather than predator birds are responsible for most of the predation on the nests of turkeys and ring-necked pheasants (Chesness et al. 1968; Hurst et al. 1996).

The relative threat posed by visual and olfactory predators should differ for canopy-nesting and ground-nesting birds, with visual predators (e.g., corvids) posing a greater threat to canopy-nesting birds and olfactory predators (e.g., mammals) posing a greater threat to ground-nesting birds. Hence, small birds that nest in forest canopies should locate their nests where visual obstructions are highest rather than where turbulence and updrafts occur; large ground-nesting birds should select nest sites based on where turbulence and updrafts will occur rather than where visual cover is greatest.

The relative threat posed by visual and olfactory predators will also vary among locations. In areas where olfactory predators predominate, animals should select hiding or nest sites based on turbulence and updrafts. In areas where visual predators pose the greater risk, hiding and nest sites should be selected based on visual cover.

Animals make decisions about where to hide at multiple landscape scales, and animals might also be able to minimize their combined risk to visual and olfactory predators by making those decisions based on whether visual or olfactory predators are more affected at a particular landscape scale. Habitat features within a meter of the nest (i.e., nest site characteristics) such as nest concealment have a major impact on the ability of visual predators to locate a nest but little impact on olfactory predators. Because the latter are influenced by atmospheric turbulence and updrafts, olfactory predators should be affected more by characteristics of habitat patch and landscape, such as the patch's roughness and aspect. This idea is supported by Badyaev (1995), who showed that in areas where only mammal predators posed a threat to nesting turkeys, nest patch characteristics had more of an influence on nests predation rates than nest site characteristics. Hence, a bird seeking a nest site that would be safe from both visual and olfactory predators may first select a habitat patch where olfactory predators will have difficulty finding its nests and then select a nest site within that patch that affords the best concealment for the nest from visual predators.

Animals can also adjust the risk to visual and olfactory predators by changing their behavior. For birds, the easiest way not to leave a depositional odor trail is to fly from place to place rather than walk (Figure 9.8). Of course, flying has its own disadvantages because it makes birds more conspicuous to visual predators. For example, a nesting duck can reduce the chances of an olfactory predator (e.g., a mammal) finding its nest by flying directly to its nest. But this same behavior will make its nest more conspicuous to a visual predator that can watch where the bird lands and then search that area for the nest. This is a common foraging practice among ravens and crows that allows them to find many nests (Hammond and Forward 1956; Erikstad et al. 1982; Hill 1984).

To reduce the chances of this happening, a bird could land away from the nest and then walk to it as cryptically as possible. Given a simultaneous threat from both visual and olfactory predators, birds flying back to their nest should not land by their nest or at a great distance from it. Rather, the optimal distance for any bird will depend on the

Figure 9.8 A bird such as this greater sage-grouse can take flight when a mammalian predator is following its depositional odor trail. (Courtesy of Justin Harrington.)

relative risk it faces from visual and olfactory predators. This risk itself will be based on the relative density of visual and olfactory predators in the area where it nests and the vulnerability of its nest to both. Consequently, birds nesting in areas where there are few avian predators or those that can defend its nest from them should land close to their nests, especially if the area has a high density of mammalian predators.

Another strategy that incubating birds can utilize to minimize their nests' vulnerability to predation is to make trips to and from the nest only in the mornings. They can then land at a distance and walk to the nest. By landing away from the nest, they minimize the risk from visual predators that hunt during the day, and by walking back in the morning, their depositional odor trail has all day to dissipate before the olfactory predators start to hunt at night. For the same reason, birds should fly directly to their nightly roosting site rather than walk so that they do not leave a fresh depositional odor trail for olfactory predators, soon to be hunting, to follow. As another example, birds that are more vulnerable to mammalian predators than avian predators while foraging on the ground should fly from one foraging patch to another so they do not leave a depositional odor trail and then concentrate their walking in a small area. In contrast, birds that are vulnerable to avian predators should walk as much as possible, providing there is enough cover to hide them while walking.

Trade-offs between the need to avoid olfactory predators and to meet the other necessities of life

To survive and reproduce, animals must complete a plethora of tasks. They must meet their nutritional needs, avoid becoming too hot or cold, and reproduce. All of these needs will affect an animal's optimal hiding strategy. For instance, if there is an ideal place to hide from predators but there are no food resources near it, then animals will not be able to use it. Likewise, they cannot remain for long in areas that are so cold or hot that they have trouble maintaining their optimal body temperature. Bare boulders on a south-facing slope may get hot enough during a hot sunny day to create such a strong updraft that an animal's odorant trail is carried above the olfactory zone of predators. Hence, it may be a good place to hide from olfactory predators, but if it is too warm, the animal may

Figure 9.9 Bobwhite quail. (Courtesy of Lee Stribling.)

overheat or have to spend so much energy panting (mammals) or gular fluttering (birds) that it cannot remain on the boulder (Sidebar 9.2).

Sidebar 9.2: Do Northern Bobwhites (*Colinus virginianus*) Select Cover Based on the Need to Avoid Predators or for Thermoregulation?

To answer this question, Guthery et al. (2005) studied the habitat selection of 217 radio-collared bobwhites (Figure 9.9) and were able to locate their birds over 9000 times. Their 800-ha study site in the Texas Panhandle contained several cover types: riparian, grassland, woodland, salt cedar, and mixed shrub. The authors report that the bobwhites showed a strong selection for mixed shrub, a weak selection for salt cedar, and an avoidance of the other cover types. Of the cover types in the area, mixed shrub provided cover from visual predators, and this savanna-like cover also caused odorant trails to disperse quickly. These same cover types also allowed bobwhites to remain in a thermo-neutral zone. Hence, Guthery et al. (2005) conclude that the need to avoid predators and the need to remain in a thermo-neutral zone led to similar predictions of bobwhite behavior and that both factors undoubtedly influence cover selection by bobwhites.

Animals not only must survive, but also must reproduce. Many animals make themselves conspicuous to potential mates and rivals in their need to acquire a territory and defend it or to find a mate. They may do so by releasing sexual pheromones, by having bright coloration, or by engaging in behaviors that call attention to themselves (for example, singing by passerines). Although these pheromones, coloration patterns, displays, and behaviors are designed to communicate with conspecifics, the information is also available to predators. Hence, predation rates on mature adults can be much higher during the breeding season than during other periods of the year.

Trade-offs between the need to reproduce this year versus during future years

Those individuals that produce the greatest number of surviving offspring will have the greatest *inclusive fitness*, and their genes will dominate in future generations, including the genes that influence their reproductive behavior and choices involving parental care. Two conflicting needs that many parents have to balance are the need to devote enough resources, such as food, to the current year's offspring so that they can survive and their own need to keep enough resources for themselves so that they will not jeopardize their opportunity to survive and to reproduce the next year. Species that reproduce only once in their lifetime can devote all of their resources and parental care to this year's offspring. For long-lived species that can reproduce over several years, their best decision may be to sacrifice one year's offspring rather than risk their own survival.

Parental care can be defined as anything that a parent does to enhance the survival of its offspring at some cost to itself. Parental care involves taking risks, including the risk of losing one's own life to a predator. These risks can be substantial. In many ground-nesting duck species, such as mallards (*Anas platyrhyynchos*), adult males outnumber adult females because so many females are killed by predators while incubating their eggs (Johnson and Sargeant 1977). Because most ducks have the ability to produce offspring for several years, predators, such as American crows, rats, opossums, and striped skunks, that commonly eat eggs or small chicks but rarely kill an adult bird are much less of a threat to a female duck's inclusive fitness than a predator (e.g., a red fox) that commonly kills incubating ducks (Figure 9.10). For this reason, ducks and other birds are more likely to select nesting sites where they are safe from red foxes and other predators that threaten their lives even if such sites increase the probability of them losing their clutches to less-dangerous predators.

Figure 9.10 Predators that commonly kill incubating ducks (such as this red fox with a mallard hen in its mouth) pose more of a threat to an adult duck's lifelong inclusive fitness than a predator that only eats eggs. (Courtesy of Alan Sargeant and the U.S.G.S. Northern Prairie Wildlife Research Center.)

Figure 9.11 Herbivores often must leave security cover and venture into open areas in search of food. (Courtesy of Mark McClure.)

Trade-offs involving the timing of dangerous activities

It is in the best interest of animals to engage in risky behavior that leaves them open to attack by predators during those periods when they are safest from predators and to remain in protective cover when predators are hunting. One of these risky behaviors is foraging. It is a risky behavior because the animal has to venture into areas to forage that it would normally avoid (Figure 9.11). For instance, American robins (*Turdus migratorius*) are safer from mammalian predators if they remain in the treetops rather than searching the ground for earthworms. Foraging is also risky because the animal's attention is distracted because it often cannot search for food and scan for predators at the same time. Thus, one would expect that animals, such as robins, would forage when predators are absent or not hunting.

Yet, there is a paradox about the timing of foraging behavior that has never been fully explained. Dawn and dusk are the best time for olfactory predators to detect and locate prey for several reasons. First, the ground is often moist so that moving animals will leave a conspicuous depositional odor trail that can be easily followed. Second, at dawn and dusk, wind velocities are often low, the atmosphere is stable, and inversions are located close to the ground. Hence, odor plumes are strongest, and olfactory predators have an easy time using them to locate prey during crepuscular periods. Yet during these periods, there also is enough light to allow visual predators to hunt.

Given that both visual and olfactory predators are hunting during these crepuscular periods, these time periods would seem to be an excellent time for any prey to remain in a secure area. Yet, many animals actively forage during these periods (Figure 9.12) (Caro 2005). How can one explain this apparent paradox of animals exposing themselves to risk when it would seem to be unnecessary? One hypothesis is the *prey availability hypothesis*. It argues that animals are forced to forage at dawn and dusk despite the risk of doing so because that is when their food is available. Although this might explain why predators and insectivorous birds forage during crepuscular hours, it would not explain why herbivores and grainivores do so because their food is available during all hours of the day and night. Hence, this hypothesis does not explain why crepuscular feeding is so ubiquitous among different types of foragers (Bednekoff and Houston 1994).

Figure 9.12 These deer are foraging at sunrise when both visual and olfactory predators are hunting. (Courtesy of Lou Cornicelli.)

A second hypothesis is the *efficient digester hypothesis*. It argues that herbivores need to digest as much food as possible during a 24-hour period and therefore to keep their digestive system filled for as long as possible (Bednekoff and Houston 1994). The problem diurnal herbivores face is that at night they may run out of food to digest. Hence, they need to eat as late in the day and as early in the morning as possible to minimize the time when their digestive system is inactive. Hence, their digestive systems drive animals to forage during crepuscular periods despite the risk. For the same reasons, herbivores, insectivores, and grainivores that forage at night would also need to be active at dawn and dusk.

The alternate hypothesis is the *vigilant prey hypothesis*. It notes that prey have developed both olfactory and visual sensory systems to avoid predators. At night, olfactory predators have an advantage over their prey because of their better olfactory ability. During the day, visual predators have an edge over their prey because of their better eyesight. This hypothesis argues that during crepuscular periods, prey can use both sensory systems and therefore have an advantage over predators that specialized in the use of one sensory system. How much safer it is to forage during twilight differs among species. Dark-eyed juncos (*Junco hyemalis*) are subject to a greater risk of predation during the dim light of dawn and forage only during these times when they can do so while remaining in cover or when starved (Lima 1988a, 1988b). In contrast, pelagic fish face a much lower risk of predation during crepuscular periods and therefore forage actively during these periods (Clark and Levy 1988).

Several predictions can be made that would be true if the efficient digester hypothesis is correct but false if the vigilant prey hypothesis is correct. First, herbivores and grainivores should forage only during crepuscular periods when forced to do so out of nutritional necessity. Hence, diurnally active prey should avoid crepuscular foraging during the summer when hours of darkness are short, and they should exhibit a high propensity for crepuscular foraging during the winter when nights are long. The opposite pattern should appear in nocturnally active herbivores and grainivores.

The second prediction is that if the efficient digester hypothesis is correct, then prey should forage only during crepuscular periods and not engage in other risky activities during these periods. This prediction, however, does not seem to be true. Many risky

Figure 9.13 Jackrabbits are more likely to nurse their young shortly after sunset than during the middle of the day or night. (Courtesy of Fred Knowlton.)

activities, such as territorial defense, aggressive interactions, and vocalizing, are more likely to occur around dawn or dusk than in the middle of the day or night. For example, jackrabbits (*Lepus* spp.) only nurse their young once or twice a day (Figure 9.13) and are more likely to do so shortly after sunset than during the middle of the night or day (F. Knowlton, personal communication, 2006).

The third prediction is that prey normally crepuscular in their foraging behavior should cease their activity during crepuscular periods if they have ample food supplies available to them during the periods when they are normally inactive. This prediction is not borne out, at least for eastern gray squirrels (*Sciurus carolinensis*), because captive squirrels that have cached food in their den so that they can feed without risk are still active at dawn and dusk (personal observation).

Trade-offs among injuries, illness, starvation, and predators

An animal that is injured, ill, or starving faces an increased risk of predation. For example, wolves and hyenas are good at picking out those individuals in a herd that are less able to escape or defend themselves because they are weak from illness or hunger. Animals that are ill or injured also are less wary, spend less time scanning the environment for predators, allow predators to get closer before fleeing, and then flee shorter distances. Deer fawns that have diarrhea or an elevated body temperature or are releasing mucus, pus, or blood are all producing more odorants than a healthy fawn. These sick or injured individuals have a higher probability of discovery and killing by coyotes than healthy ones (Cook et al. 1971).

Furthermore, fawns that are abandoned by their mothers or do not receive adequate quantities of milk are more likely to wander about and vocalize (Linsdale and Tomich 1953; Cook et al. 1971). Such behavior increases the conspicuousness of fawns to both visual and olfactory predators and leads to higher rates of predation. When Gonzales County, Texas, experienced a drought during 1971, most fawns died, but almost all were

Figure 9.14 During years when prey populations crash, both ferruginous hawks are forced to leave their nestlings undefended while both parents are simultaneously hunting for food. Because of this, nestlings are more likely to be killed by a predator than starve to death during a famine. (Courtesy of Heather Keough.)

killed by predators rather than starving to death (Carroll and Brown 1977). Likewise, gull chicks are more likely to abandon their nest and leave their parents' territory when they are hungry. Such chicks have a short life expectancy because they are killed by neighboring gulls.

During famines, most ferruginous hawk (*Buteo regalis*) chicks die, but surprisingly most are killed by predators rather than by starving to death (Keough 2006). Apparently, there are two reasons for this pattern. First, both parents spend more time away from the nest looking for food and leave the young at the nest unprotected (Figure 9.14). Second, other raptor species, such as golden eagles (*Aquila chrysaetos*), are also stressed by the food shortage and are more willing to prey on ferruginous hawk chicks despite the risk of attack by the parents (Keough 2006).

Similarly, predation rates decrease when food supplies are abundant. Ward and Kennedy (1996) demonstrated this in an elegant experiment by providing supplemental food to some northern goshawks (*Accipiter gentilis*) but not others (i.e., control hawks). Their results demonstrated that supplementally fed hawks had higher reproductive rates than control hawks. However, the reason for this was because the supplementally fed hawks spent more time near their nests, and fewer of their young were killed by predators.

Inadequate food supplies also force animals to forage in areas where the risk of predation is high. Burrowing animals, such as prairie dogs, chipmunks, and pika (*Ochotona princeps*), travel greater distances from the safety of their burrows when closer food supplies have been consumed (Figure 9.15). Other herbivores (for example, ungulates and jackrabbits) forage in risky habitats that they would normally avoid when droughts reduce plant production.

The opposite also occurs. An increased threat of predation can force animals to forage in large groups despite increased competition for food from other members of the group or to stop foraging in profitable areas where they are particularly vulnerable to predators (Figure 9.16). For instance, the introduction of wolves into the Yellowstone ecosystem caused elk to change both their foraging areas and foraging behavior and increased their risk of starvation (Mao et al. 2005). Thus, there is a complex interplay among predation, illness, injury, and starvation. Often, a combination of factors leads to an animal's death.

(A)

(B)

Figure 9.15 Pikas (A) and chipmunks (B) travel greater distances from the safety of their burrows when food supplies are limited. (Courtesy of Madsen.)

Figure 9.16 During winter, mule deer forage in large groups in areas where there are cougars (*Felis concolor*) or wolves. (Courtesy of Christine Peterson.)

Summary

The great conundrum for the theory of optimal hiding strategies is why there is such intraspecific variation in nest site selection and optimal hiding strategies. It has been assumed that this variability results because some individuals have inferior knowledge (Bekoff et al. 1989) or are forced to nest in poor sites because optimal nest sites are limited (Norment 1993). In some cases, this variability occurs because future nesting conditions are unpredictable (Leonard and Picman 1987).

The evidence in this chapter indicates that this high level of individual variation results because birds make countless trade-offs in selecting a nest site, and that they use the experiences gained not only during prior nesting attempts, but also over their entire lifetime to try to make the correct decisions. Even then, predators have an advantage over their prey because they can refine their optimal search image faster than animals can refine their optimal hiding strategy. Hence, birds are trapped in a lose-lose situation when trying to select safe nest sites, and they make the best of a bad situation based on their imperfect knowledge.

chapter ten

Impact of olfactory predators on the behavior of female ungulates during parturition and on the behavior of their young

Female *ungulates* (herbivorous mammals with hooves) have difficulty defending themselves against mammalian predators while giving birth (*parturition*). Likewise, fawns and calves are vulnerable from their birth until they are old enough to defend themselves or outrun predators. Many mammalian predators, including black bears (*Ursus americanus*), grizzly bears (*Ursus arctos*), coyotes, and red foxes, specifically hunt for ungulate young, and mammalian predation is the leading cause of death for many young ungulate species (Hirth 2000). To locate these neonates, predators such as bears develop a search image of areas where fawns and calves are likely to be hiding based on experiences gained during prior hunting trips (Figure 10.1). They then use both their vision and olfaction to locate the hiding young.

To survive, young ungulates must avoid detection by predators. As is true in most situations, humans are aware of the cryptic nature of fawns and calves and how hard they are to locate visually. We are less aware of the methods used by these young to avoid

Figure 10.1 Bears use both their visual and olfactory senses to hunt for ungulate fawns and calves. (Courtesy of U.S. Department of Agriculture's Wildlife Services.)

detection through olfaction. Likewise, many scientists have examined how neonates avoid detection by visual predators, but few studies have examined how they avoid detection by olfactory predators (Caro 2005). As we have discussed in this book, olfactory predators have difficulty finding prey when atmospheric turbulence and updrafts are present, when prey produce few odorants, or when atmospheric conditions are unfavorable. This chapter examines whether young fawns, calves, and their mothers take advantage of these locations and conditions to lessen the probability of detection by olfactory predators.

As one example, updrafts should be more common on south-facing slopes than north-facing ones because the former will be warmer and will therefore experience more updrafts. Hence, olfactory predators will have a harder time locating prey on a south-facing slope. But, do ungulates use these slopes for parturition or to hide their neonates? The scientific literature indicates that several species do so. In California, mule deer fawns prefer to bed down on south-facing slopes (Bowyer et al. 1998). Bergerud et al. (1984) report that caribou seek out south-facing slopes to give birth, and caribou calves hide there during their first days of life. This may be because south-facing slopes are warmer as it is cold where caribou live. However, elk (*Cervus canadensis*) fawns select south-facing slopes for bedding in many parts of their range, including Arizona, where low temperatures are not a problem (Houston 1982; Wallace and Krausman 1992; Slovkin et al. 2002). Hence, there may be more to why ungulates select south-facing slopes than just thermoregulation.

Ungulate species use two basic strategies to avoid predation on their young. In the *follow-mother strategy*, fawns stay close to their mothers and follow them as they travel. For this strategy, mothers or their kin protect the young by warding off predators, or the mothers and their young join a large herd and seek safety in numbers (Lent 1974). Wildebeest (*Connochaetes taurinus*) and American bison (*Bison bison*) are good examples of species that utilize the follow-mother strategy. In the *hiding strategy*, young remain by themselves and hide from predators. This strategy is followed by those species with altricial young, such as white-tailed deer, mule deer (*Odocoileus hemionus*), pronghorn (*Antilocapra americana*), and elk (Lent 1974; Hirth 2000). To escape attention, hiders are camouflaged with spotted or striped patterns (Figure 10.2) and lie on the ground motionless for long periods of time. When alarmed, they freeze and rarely move even when a predator is within a few meters of them.

Figure 10.2 Young fawns (this one is a mule deer) follow a hiding strategy to avoid predators and have a spotted coat for camouflage. (Courtesy of Justin Harrington.)

Table 10.1 Age of Ungulate Fawns and Calves When They Stop Their Constant Hiding to Avoid Predation and Start Following Their Mothers

Species	Age (weeks)
Siberian ibex (*Capra ibex siberica*)	2–3
Mouflon sheep (*Ovis musimon*)	3
Red deer	3–4
Feral goat (*Capra hircus*)	4
Axis deer (*Axis axis*)	7–10
Elk	18–20
Muntjac (*Muntiacus muntjak*)	14–21
Roe deer (*Capreolus capreolus*)	14–21
Pronghorn	14–21
Defassa waterbuck (*Kobus ellipsiprymnus defassa*)	14–28
Klipspringer (*Oreotragus oreotragus*)	28
Dorcas gazelle (*Gazella dorcas*)	14–42
Thomson's gazelle (*Gazella thomsoni*)	14–42
Indian blackbuck (*Antilope cervicapra*)	14–42
Kudu (*Tragelaphus* spp.)	21–49
Uganda kob (*Adenota kob thomasi*)	56–112
Reedbuck (*Redunca* spp.)	56–112

Source: Based on data from Lent, P.C., in V. Geist and F. Walther, Eds., *Symposium on the Behaviour of Ungulates and Its Relation to Management,* IUCN Publications 24, Morges, Switzerland, 1974, pp. 14–55.

Figure 10.3 Once mule deer are a few months old, as are the three in the center of this photo, they start following their mothers and rely on their speed to avoid predators. (Courtesy of Justin Harrington.)

However, the differences between followers and hiders are relative rather than absolute. Rather, the duration of this hiding phase varies among species (Table 10.1) and usually ends when the young have gained sufficient agility and speed so that it has a better chance of avoiding killing by a predator by fleeing rather than hiding (Figure 10.3). Hence, in examining how young avoid detection by olfactory predators, I use data from species with both hiders and followers but am only concerned with that period before the young are able to follow their mothers.

Do females reduce their production of odorants at parturition sites or the bedding sites of their young?

When ungulate females are about to give birth, they leave their herd or social group and seek a spot where they can be by themselves. Some ungulates, such as white-tailed deer, mule deer, and roe deer, defend their *parturition sites* against other deer and drive away even their own kin (Ozoga et al. 1982 and references therein). This quest for isolation certainly has the outcome of reducing the amount of odorants in the parturition area by reducing the number of other ungulates in the area.

After giving birth, females consume the placenta, amniotic membranes, and any blood or bodily fluids; this occurs in many ungulate species, including pronghorn, red deer, gazelles (*Gazella* spp.), kudu, sitatunga (*Tragelaphus spekii*), white-tailed deer, mule deer, Uganda kob, Barbary sheep (*Ammotragus lervia*), and caribou (Autenrieth and Fichter 1975). This behavior reduces the predation risk to neonates by reducing odorant concentrations at parturition sites.

Mothers also start licking their neonates within a few minutes of birth (Figure 10.4). Licking usually covers the entire body and is concentrated on the head and anogenital region. Mothers also lick objects in the environment and consume soil contaminated with the neonate fetal membranes and urine (Lent 1974; Autenrieth and Fichter 1975; Geist 2002). Many authors have argued that licking helps dry the neonate, plays an important role in establishing a mother-young bond, and helps mothers learn to recognize their own young by smell (Lent 1974). However, mothers continue to groom their young for several days after their birth. Once neonates begin to defecate, mothers also eat the feces and soil contaminated with feces or urine (Gilbert 1972; Lent 1974; Geist 2002). Although females may receive some nutritional benefits from these behaviors, they also reduce odorants that olfactory predators could use to locate their young. Such behavior must be quite successful in reducing odorants. Many authors have reported that neonates lack any odor (e.g., Johnson 1951; Mech 1984; Linnell et al. 2004), and there are many anecdotal reports of hunting dogs standing above or passing by an ungulate fawn or calf without realizing it (Johnson 1951; Lent 1974).

Figure 10.4 This pronghorn fawn is just a few hours old but has already been licked several times by its mother. (Courtesy of Madsen.)

Figure 10.5 Females produce large amounts of milk so they can feed their young infrequently and reduce the probability of olfactory predators following their depositional odor trail to their fawns. (Courtesy of Justin Harrington.)

Although it is impossible for a neonate not to produce some odorants (e.g., metabolic odorants), both the infant and mother appear to keep odorants released by a fawn or calf to a minimum. For instance, in species with hiding young, mothers keep an eye on their young from a considerable distance. Normal distances of mothers from their hiding young are 30 to 300 m for elk, more than 500 m for Grant's gazelle (*Gazella granti*) and feral goats, and 500 to 1000 m for pronghorn and Defassa waterbuck (Lent 1974). Feedings are kept to a minimum, and when they do occur, copious amounts of milk are provided (Figure 10.5, Table 10.2).

Females also do not approach their young when there are mammalian predators in the vicinity (Caro 2005). Female ungulates do not walk up to their hiding young (Figure 10.6). Rather, the mothers stand 10 to 50 m away and wait for the infant to approach them (Lent 1974). By maintaining their distance from their young and feeding them infrequently, mothers help protect their young from olfactory predators because adults produce more odorants than infants; this same tactic is used to protect young from visual predators that might be watching the mothers to learn where their young are located. Mothers also frequently move their young to new hiding areas (Caro 2005). Although this behavior may expose them to visual predators, it reduces their vulnerability to olfactory predators because odorants at bed sites build up over time.

Is the behavior of neonates designed to hinder the ability of predators to find them using olfaction?

The behavior of neonates is also designed to keep their odorants to a minimum. The young lie curled up on the ground with their head against their body and tail and legs tucked beneath or beside them (Figure 10.7). They often lay their neck and head on the ground. When disturbed their ears are tucked back against the head (White et al. 1972). This

Table 10.2 Frequency of Nursing among Ungulate Species

Species	Nursings per day
Bovidae	
Reedbuck	1–2
Defassa waterbuck	1–2
Uganda kob	2
Red lechwe (*Kobus leche*)	2–3
Sitatunga	2–4
Lesser kudu (*Tragelaphus imberbis*)	2–3
American bison	3–4
Greater kudu (*Tragelaphus strepsiceros*)	3–5
Gazelles	3–5
Dik-dik (*Madoqua* spp.)	4
Muskox (*Ovibos moschatus*)	6–8
Cervidae	
Pudu (*Pudu pudu*)	3
Red deer	6
Roe deer	6–7
Caribou	>10

Source: Based on data from Lent, P.C., in V. Geist and F. Walther, Eds., *Symposium on the Behaviour of Ungulates and Its Relation to Management*, IUCN Publications 24, Morges, Switzerland, 1974, pp. 14–55.

Figure 10.6 Females do not visit the bed sites of their fawns but rather keep their distance until the fawns approach them. (Courtesy of U.S. Department of Agriculture's Wildlife Services.)

position minimizes the amount of their body surface that is in contact with the air and therefore minimizes their emission of odorants. It also reduces their silhouette so that it is harder for predators to recognize them visually. When scared, the young freeze into this position and will not move even when a predator is within a few meters of them (Table 10.3). This freezing response seems to be widespread among Artiodactyla species, such as white-tailed deer, mule deer, red deer, pronghorn, Grant's gazelle, Thompson's gazelle, caribou, elk, moose, and American bison (Lent 1974; Autenrieth and Fichter 1975; Geist 2002). Fawns of white-tailed deer and red deer slow their heart rates (i.e., bradycardia) and respiration rates by as much as 40% when alarmed (Espmark and Langvatn 1979,

Figure 10.7 When alarmed, young fawns will curl up into a ball with legs tucked beneath them. Doing so reduces their odor plume by diminishing their surface area in contact with the air. (Courtesy of Russ Kinne and Tony DiNicola.)

Table 10.3 Age (days) When Ungulate Fawns and Calves Stop Freezing into a Prone Position Whenever They Are Approached by a Predator or Are Alarmed

Species	Age (days)
Fallow deer	1–2
Caribou	1–2
Mouflon sheep	3–4
Red deer	3–4
Moose	5–7
White-tailed deer	10
Roe deer	14
Greater kudu	14
Dorcas gazelle	14

Source: Based on data from Lent, P.C., in V. Geist and F. Walther, Eds., *Symposium on the Behaviour of Ungulates and Its Relation to Management*, IUCN Publications 24, Morges, Switzerland, 1974, pp. 14–55.

1985; Jacobsen 1979). Interestingly, a wolf's howl (Figure 10.8) will cause white-tailed deer fawns younger than 45 days to exhibit bradycardia (Moen et al. 1978).

In most species, calves or fawns select their own bedding sites, and twins seek separate beds, which are often 50 to 200 m apart (Lent 1974). Bedding separately reduces the amount of odorants released at each bedding site and therefore reduces the probability that an olfactory predator will find either twin.

Figure 10.8 The sound of wolves howling produces bradycardia in fawns. (Photo courtesy of U.S. Department of Agriculture's Wildlife Services.)

Do fawns adjust the timing of their movements to avoid attracting the attention of visual or olfactory predators?

Mammalian predators are responsible for most fawn mortalities in white-tailed deer (Cook et al. 1971; Carroll and Brown 1977; Nelson and Woolf 1987). Hence, the behavior of fawns should be shaped by evolutionary pressure more to avoid predation by olfactory predators than by visual predators. Hence, it is not surprising that white-tailed deer fawns move little during nocturnal hours when they would be vulnerable to olfactory predators. Instead, they are likely to move or change locations during crepuscular and daylight hours (Jackson et al. 1972; Huegel 1985; Schwede et al. 1992) even though visual predators such as golden eagles (Figure 10.9) that prey on fawns are hunting during these periods.

Do female ungulates select parturition sites, and do young select bedding grounds where olfactory predators would have a hard time finding them?

It is well-known that fawns and calves seek bedding sites where they are hidden from view. But, do female ungulates also seek parturition sites and fawns and calves seek bedding sites where they are hidden from olfactory predators? To escape detection by olfactory predators, I would predict that bedding sites would be located in habitat patches that are rough (that is, have a higher d and z_0). In mountainous areas, they should also be located on south-facing slopes and along the tops of hills and ridges. When the characteristics of bedding sites have been compared to random sites in the same area, most studies support these predictions. Some examples follow for different ungulate species.

Pronghorn

In the sagebrush-steppe habitat of Wyoming, fawn birth sites and bed sites of pronghorn (Figure 10.10) were more likely than random sites to be within a meter of a shrub cover, to have more shrub canopy cover, and to be beside a big sagebrush (*Artemisia tridentata*).

Figure 10.9 Golden eagles use their vision to hunt for deer fawns during daylight hours. (Courtesy of Justin Harrington.)

Figure 10.10 Pronghorn. (Courtesy of Justin Harrington.)

Bedding sites were also more likely to be in a habitat patch with a higher shrub canopy cover than random sites (Alldredge et al. 1991).

In the grasslands of Alberta where sagebrush was sparse, pronghorn fawns selected bed sites that were on slopes or tops of hills, in small depressions, and adjacent to clumps of silver sagebrush (*Artemisia cana*), rocks, or cow dung that provided both vertical and horizontal cover. Surviving pronghorn were more likely to have bedded in depressions and in areas with greater cover density (stands of grass or forbs more than 0.25 m in height) than nonsurvivors (Barrett 1981). Wind velocities at a height of 1.5 m were 60% less over bed sites than random sites (Barrett 1981).

In Texas, pronghorn fawns selected bed sites that were close to plants or objects that provided vertical structure and tended to bed down with their backs against the cover. Fawn bed sites had more canopy cover, grass cover, and rock cover than random sites. Hence, it is not surprising that fawn visibility was less at bed sites than random sites. Interestingly, they were also located in habitat patches with lower shrub cover and shrub densities so that visibility was greater at long ranges (>50 m). Random and bed sites were similar in slope, vegetation height, and forb cover, suggesting a lack of preference for rough habitat (Cannon and Bryant 1997). However, the bed sites of surviving fawns differed from those that died in that the former had more shrub cover, had less rocky cover, and were on flatter ground (Cannon and Bryant 1997).

Roe deer

In the agricultural regions of Norway, red foxes are the major predators of roe deer fawns. In the early spring, before grass and forbs have grown tall enough to provide much cover, more fawn beds were located in coniferous and deciduous forests than in moorland and meadows (Linnell et al. 2004). Later in the season, fawns prefer to bed in abandoned pastures where the grass and forbs are over their heads, and they avoid pastures with short grass (Linnell et al. 2004). At most bed sites, the vegetation is thick and high enough that a red fox would have to be within a meter of the fawn to see it. Elsewhere, roe deer fawns prefer bedding sites surrounded by thick vegetation (Lent 1974). Such sites would also experience higher rates of turbulence.

White-tailed deer

Throughout much of their range, predation by coyotes, black bears, domestic dogs, and bobcats is the major cause of mortality among the fawns of white-tailed deer (Figure 10.11). In some areas, mammalian predation rates on fawns approached 50% (Cook et al. 1971; Nelson and Mech 1986; Kunkel and Mech 1994; Ballard et al. 1999), but the primary predator varied among areas. In Texas and New Brunswick, coyotes were the major predator of fawns; wolves and black bears were the main threat in Minnesota.

While young, white-tailed deer fawns rely on the hiding strategy to avoid predators and select bed sites carefully. In Iowa, fawns of white-tailed deer selected bed sites irrespective of plant species but rather chose sites with vertical and horizontal cover. Cover provided by short grass, tall forb, shrubs, and saplings was greater at bed sites than at sites 5 to 10 m away from them, as was tree canopy cover (Huegel et al. 1986).

In the Black Hills of South Dakota, bed sites of white-tailed deer fawns were usually located in open stands of ponderosa pine (*Pinus ponderosa*). Bed sites differed from random sites in having more vertical and horizontal structure. Bed sites had denser vegetation cover that was dominated by grass cover and avoided shrub cover (Uresk et al. 1999). Early in the season, before herbaceous cover and grass had time to grow, fawns preferred to bed near fallen trees and debris. Interestingly, bed sites and random sites did not differ in visual obscurity. Fawns selected bed sites with a reduced tree canopy cover and hence were often found in pine stands that had been recently thinned. At these sites, sweeps and ejections are more likely to occur along the forest floor.

Mule deer

Mule deer spend much of their time on the tops of hills or high on slopes. In such locations, the deer were less likely to be detected by coyotes moving beneath them and to be attacked

Figure 10.11 White-tailed deer. (Courtesy of Russ Kinne and Tony DiNicola.)

Figure 10.12 Mule deer. (Courtesy of U.S. Department of Agriculture's Wildlife Services.)

(Lingle 2002). As pointed out in Chapter 6, the odor plumes of deer located on hilltops or high on slopes are rarely detectable to predators beneath them.

Before parturition, female mule deer (Figure 10.12) seek sites where they will be safe from mammalian predators. For instance, Steigers and Flinders (1980) observe that mule deer swam to barren islands to give birth even though these islands lacked forage.

Figure 10.13 Elk.

Mule deer fawns hide on steep slopes of greater than 30° in Arizona, and when bed sites were on flat terrain, they were located along gullies. Hence, both types of locations would be likely to experience drainage winds. Fawn bed sites were more likely than random sites to be located near palo verde (*Circidium* spp.) trees and surface features that would obscure the visibility of the fawn and cause atmospheric turbulence (Fox and Krausman 1994).

Elk

For parturition, elk (Figure 10.13) used stands of timber or heavy brush (Batchelor 1965; Wallace and Krausman 1992), sagebrush habitat (Houston 1982; Slovkin et al. 2002), or the ecotone between open prairie and forests (Johnson 1951; Harper et al. 1967). They also used sagebrush intermixed with confers (Slovkin et al. 2002). The common denominator in all of these preferred habitats is the presence of cover that helps keep the female and her calf hidden from both visual and olfactory predators. In heavily timbered areas, elk preferred to calve in forest clearings or in open forests (Slovkin et al. 2002). These are the areas where sweeps and ejections are likely to occur. South-facing slopes are preferred by elk for calving (Waldrip and Shaw 1979; Slovkin et al. 2002). This preference would help hide them from olfactory predators because updrafts are more frequent along south-facing slopes.

After birth, elk calves commonly hid beside surface features such as boulders, slash piles, fallen trees, and standing timber, which create turbulence (Waldrip and Shaw 1979; Slovkin 1982; Wallace and Krausman 1992). Even when bedding in forests, calves are usually found within 10 m of the forest's ecotone. Many authors report that elk calves hid in sagebrush and on slopes (summarized by Slovkin et al. 2002), especially on long ones (Johnson 1951). Phillips (1974) found them on slopes with grades averaging 35%.

After the arrival of wolves to Yellowstone National Park, elk shifted their habitat preference to forests that had burned in prior years. Burned forests had more open canopies, more subcanopy vegetation, and more fallen trees than unburned forests (Mao et al. 2005). Hence, the forest floor of burned forests had a higher z_0 and d than unburned forests and contained more surface features within the wolf's olfactory zone. For all of these reasons, wolves should have a harder time locating prey in burned forests than in unburned ones. Moreover, wolves spent less time hunting in burned forests despite the fact that elk preferred to hide in such areas (Mao et al. 2005). In addition, the arrival of wolves caused elk in Yellowstone National Park to spend more time in grassland and less time in shrub land (Mao et al. 2005). Elk also selected steeper slopes in areas where wolves occur compared to where they do not. Furthermore, wolf kills were more likely to occur on level ground and shallower slopes than on steep slopes.

Caribou

Pregnant caribou (Figure 10.14) leave the herds and move to traditional calving areas that have been used in previous years, and once there, the pregnant females space themselves out (Skogland 1989). In both northern Europe (Skogland 1989) and North America (Bergerud et al. 1984), these calving areas are usually high in the alpine zone and usually consist of rugged areas with a broken relief (that is, high densities of hills, ravines, cliffs, boulders, and other vertical surface features). Calving areas often consist of a mixture of snow-covered ground and bare areas. Such areas would reduce the ability of predators to follow odor plumes to the calves. The broken landscape would maximize mechanical turbulence, and the mixture of snow and bare areas in close proximity would increase convective currents and cause odor plumes to rise above the olfactory zone of the calves' main predators, including grizzly bears, wolverines (*Gulo gulo*), and wolves. This would be especially true considering that caribou calves normally hide on the bare ground, which would be warmer than snow drifts.

Bergerud et al. (1984) noted that caribou climb into the high mountains of British Columbia, Canada, to give birth on south-facing slopes. At Spatsizi Provincial Park, they noted that wind velocities are higher at the calving grounds than at lower elevations where the caribou normally forage. They hypothesized that doing so prevents their odor plumes

Figure 10.14 Caribou. (Courtesy of James Schaefer.)

from reaching predators at lower elevations because the wind usually blows from the south in the park.

Woodland caribou persist in Ontario, Canada, primarily by occupying the shorelines of Lake Superior and other large lakes during the winter and use offshore islands for parturition sites (Bergerud 1985). A major advantage of existence along shorelines for caribou is that they can take to the water if chased by wolves. Another advantage is that land breezes created during nights when the water is warmer than the ground surface would carry the odor plumes of the caribou out over the lake and away from wolves. This benefit would be greatest in late fall and winter when Lake Superior is warmer than the mainland.

Dall's sheep and bighorn sheep

When compared to random sites, habitats used for lambing by Dall's sheep (*Ovis dalli*) were more likely to be steeper, snow free, and south facing (Rachlow and Bowyer 1998). They also had more grass cover and a rougher surface (that is, surface features were more prominent).

Pregnant bighorn sheep (*Ovis canadensis*) (Figure 10.15) moved to more precipitous and rugged habitat for parturition (Festa-Bianchet 1988; Berger 1991; Bleich 1999). They did not do so for nutritional reasons because they often left areas with high-quality forage and moved to areas where forage was scarce and of low quality (Festa-Bianchet 1988). When Krausman and Shackleton (1999) compared parturition sites used by desert bighorn

Figure 10.15 Big-horned sheep. (Courtesy of Madsen.)

sheep to sites used by females during the rest of the year, they found that parturition sites were steeper and more rugged. Females with lambs occupied steeper slopes and rougher habitat than males or females without lambs (Bleich et al. 1997). Furthermore, habitat ruggedness increased the probability of lamb survival (Berger 1991; Bleich 1999).

chapter eleven

Do nest site characteristics influence nest predation rates by olfactory predators?

In this book, I have shown that suboptimal wind speeds, updrafts, and turbulence reduce the ability of olfactory predators to locate prey, and that these atmospheric conditions are predictable in both time and space. In this and the following chapters, I examine whether nesting birds take advantage of these atmospheric conditions by locating their nests in areas where these conditions occur and whether those birds that do so experience a lower risk of predation from olfactory predators than birds that hide elsewhere. Evolutionary forces should favor birds that can nest in safe locations because predation is responsible for most (80%) nest failures (Martin 1993; Chalfoun et al. 2002a; Chalfoun et al. 2002b).

I test the olfactory concealment theory by using it to generate predictions about where birds should hide their nests. I then use the scientific literature to determine if the predictions are true. Some of these predictions are accurate, some are not, and some have not been tested.

Impact of avian mass, surface area, and metabolic rates on olfactory predators

Prediction 11.1

> Larger birds should release more odorants than smaller birds, and olfactory predators should have an easier time finding their nests. Therefore, the nests of large birds should be more vulnerable to olfactory predators than those of small birds.

Metabolism and body surface area are positively correlated with body mass. Therefore, both production and emission of odorants should also be positively correlated to body mass regardless of whether they are metabolic odorants or surface odorants. Because of this, olfactory predators should be able to detect and locate the nest of a large bird more easily than a small one (Figure 11.1). When this prediction was examined using birds nesting in riparian habitats in Iowa, however, the rate of nest predation by midsize mammals (for example, raccoons, striped skunks, opossums, coyotes, and domestic dogs) decreased with increasing bird mass and was actually highest on birds weighing less than 20 g (Best and Stauffer 1980). Such a pattern might be expected among cavity-nesting

Figure 11.1 Large birds, such as these double-crested cormorants (*Phalacrocorax auritus*), and their young produce great amounts of odorants. These odorants make it easy for an olfactory predator to locate their nests. (Courtesy of U.S. Fish and Wildlife Service's Bear River Migratory Bird Refuge.)

species because competition among birds for cavities is keen, and bigger birds can force smaller ones to use cavities that are more vulnerable to predators, such as those closer to the ground (Van Balen et al. 1982; Nilsson 1984).

However, few of the species examined by Best and Stauffer (1980) were cavity nesters. Instead, Best and Stauffer (1980) believe that larger birds suffered less nest predation because they could defend their nests from large mammalian predators better than smaller birds could. Yet, the only birds in their study that weighed more than 80 g were mourning doves (*Zenaida macroura*). Such small birds are physically incapable of defending their nest against a midsize mammal. Hence, I cannot explain the results of Best and Stauffer (1980) based on the olfactory concealment theory.

> Prediction 11.2
>
> When incubating birds or nestlings detect the presence of an olfactory predator, they should lie down as low as possible so that the smallest amount of their body surface is exposed to the air. Doing so reduces the amount of odorants released by their body surfaces.

The quantity of odorants emitted by an animal's surface depends in part on the amount of its surface area exposed to the air. For this reason, incubating birds and hatchlings should lie as low as possible (that is, keep as much of their body surface in contact with the nest) when they detect the presence of an olfactory predator. This behavior is commonly exhibited by both incubating birds and nestlings when startled. This behavior also conceals birds from visual predators by reducing their profile so that they will be less conspicuous (Caro 2005).

Prediction 11.3

> When incubating birds or nestlings detect the presence of a predator,
> they should reduce their metabolic rate or hold their breath as long
> as possible to disrupt the emission of metabolic odorants.

The respiration rate of a bird when incubating is less than its rate when perching (Caro 2005); a low respiration rate makes it more difficult for olfactory predators to locate nests by detecting metabolic odorants. Furthermore, some bird species, such as common poorwill (*Phalaenoptilus nuttallii*), goatsuckers, hummingbirds, and swifts, that roost in cavities through the winter exhibit torpor and do not maintain their normal body temperatures. When torpid, a common poorwill can reduce its body temperature to 5°C and its metabolic rate by 97% (Welty 1962). Although the main benefit of torpidity is to allow these birds to conserve energy during the winter, this behavior also has the benefit that it reduces their output of metabolic odorants so that olfactory predators are less likely to find them when they are torpid and helpless.

Another important question is whether incubating females can quickly reduce their respiration and heart rates when a predator is near. The answer, in some cases, is yes. Gabrielsen et al. (1985) monitored the heart and respiration rates of incubating willow ptarmigan (*Lagopus lagopus*) and Svalbard ptarmigan (*Lagopus mutus*). They found that when a predator was approaching, ptarmigan hens reduced their heart rates from 204 to 119 beats per minute and their respiration rates from 25 to 12 breaths per minute.

When capture is imminent and escape impossible, many species of reptiles, amphibians, mammals, and birds feign death (Figure 11.2). Although their eyes are open and they are aware of their surroundings and opportunities to escape, animals that feign death are in a catatonic-like paralysis during which they remain motionless and their cardiac and respiratory rates decline (Gabrielsen et al. 1985; Caro 2005).

Death feigning can be considered a last-chance effort to avoid being killed by a predator by denying it the stimuli needed to elicit killing behavior, including the release of metabolic odorants. It can be effective in some cases. In one study using captive animals,

Figure 11.2 This opossum is "playing opossum," meaning that it seems dead but it is really in a catatonic state with a reduced heart and respiration rate. (Courtesy of Madsen.)

29 of 50 ducks that feigned death when seized by red foxes escaped with their lives (Sargeant and Eberhardt 1975). Not surprisingly, death feigning works best against young, naïve foxes.

Impact of nest characteristics on olfactory predators

Prediction 11.4

> Incubating birds and nestlings inside dome nests or open-cup nests with a deep pocket and high sides will release fewer odorants from their body surface than those in shallow-cup nests. Hence, dome nests are less likely to be depredated by olfactory predators than are cup nests.

The number of odorants emitted by an animal's surface will depend on the amount of its surface area exposed to the air. For this reason, incubating birds and nestlings should release fewer surface odorants and be less vulnerable to olfactory predators when in dome nests or open-cup nests with high sides. This prediction is accurate; numerous studies have demonstrated that birds that build dome nests suffer less nest predation than birds nesting in open-cup nests (Hansell 2000; Caro 2005).

Prediction 11.5

> Incubating birds or hatchlings in nests that are dense, tightly woven, and less porous will release fewer odorants into the air than those in more porous nests. These denser nests therefore are less likely to be depredated by olfactory predators than porous nests.

All nests contain air pockets through which odorants can pass, with the porosity of nests to odorants varying with the amount of air contained within them. Porosity can be determined by measuring the volume of a nest and the volume of the nesting material after it has been dried and compressed into a solid pellet. By subtracting the latter from the former, the volume of air in a nest can be determined. The ratio of air volume to total nest volume provides a measure of the porosity of a nest to odorants. Porous nests will provide incubating birds and nestlings less protection from olfactory predators than less-porous nests. Mertens (1977) used this approach to determine that the porosity of great tit (*Parus major*) nests was 0.98 (proportion of the nest's volume that is composed of air) but did not compare nest porosity to the vulnerability of nests to olfactory predators.

There are at least a few studies that support the prediction that birds in porous nests are more vulnerable to olfactory predators than those in nonporous nests. Rangen et al. (2000) compared predation rates on eggs placed in wicker nests that had been camouflaged by sewing camouflaged cloth on the outside to break up the nest's outline to eggs placed in wicker nests that had been caked in mud. The mud nests were less porous than the others. They found that avian predators did not discriminate between the two nest types, whereas mammalian predators that used olfaction to locate nests depredated camouflaged nests more than mud nests.

Martin (1987) compared predation rates on eggs placed in (1) real MacGillivray warbler (*Oporornis tolmiei*) and hermit thrush (*Catharus guttatus*) nests saved from the prior year, (2) wicker nests lined on both the outside and inside with moss to camouflage them and make them appear more like a real MacGillivray warbler nest, and (3) wicker nests lined only on the inside with leaves. To his surprise, he found that leaf-lined nests were depredated less than the other two nest types, although the former were the most conspicuous and easiest to see, at least to human eyes. One explanation for Martin's finding is that an egg inside a leaf-lined nest would have less of its surface area exposed to the air than an egg in a moss-lined nest and therefore would be less conspicuous to an olfactory predator. It is worth noting that, at his study site, mammalian predators were much more numerous than visual predators (Martin 1987).

Baya weaver (*Ploceus philippinus*) males build the nest by themselves and then use the nest to attract a female mate. From the standpoint of trying to increase her inclusive fitness, a female should select that nest that offers the greatest benefit to herself and her young. In Baya weavers, only 23% of nesting attempts successfully fledged young, and most attempts failed because of rodent and snake predation (Quader 2006). The one attribute of nest construction and architecture that was important to females in selecting a nest was whether the nest was tightly woven with fine fiber. Such nests also suffered lower nest predation rates (Quader 2006). These tightly woven nests were better insulated than loosely woven nests (Quader 2006), and tightly woven nests release fewer odorants and do so in more intermittently than loosely constructed nests. Hence, both snakes and rodents have a harder time following the odor plume back to tightly woven nests.

Prediction 11.6

> Incubating birds and nestlings in dry nests will have lower metabolic rates and release fewer metabolic odorants than those in wet nests. Hence, dry nests should experience lower predation rates from olfactory predators than wet ones.

The heat conductivity of a nest depends on its thickness and the proportions of its volume composed of water, air, and solid material. The heat conductance ($Wm^{-1}C^{-1}$) of water is 0.6, which far exceeds that of air (0.024) and dry nesting material (0.1 to 0.15) (Mertens 1977). Thus, it is mainly the water content of a nest that determines its ability to conduct heat. Incubating birds and nestlings have the highest metabolism and release the most metabolic odorants on nights when their nests are saturated with moisture. Hence, nest predation rates by olfactory predators should be higher on wet nests than dry ones.

I am unaware of any studies that have tested this prediction directly. However, Keith (1961), Livezey (1981), and Crabtree et al. (1989) reported that ground-nesting ducks suffered higher rates of nest predation when their nests were close to water then when their nests were farther away from water. In their review of the ornithological literature, Batary and Baldi (2004) found the same phenomenon was true for other birds as well. Keith (1961) and Townsend (1966) note that nests near water were built on moist soil and speculate that the moisture around these nests enhanced odors emanating from the nest and made the nests easier to detect.

Prediction 11.7

> Cavity nests should emit fewer odorants into the air and should do so more intermittently than open-cup nests. Therefore, olfactory predators should have a harder time locating cavity nests than open-cup nests, and predation rates by olfactory predators should be lower on cavity nests than on open-cup nests.

Cavity nests often suffer less predation than open-cup nests (Oniki 1979; Wilcove 1985; Nilsson 1986; Caro 2005), and there are three hypotheses that can explain this. First, some predators may lack the dexterity to reach eggs inside a cavity. Second, cavity nests may be harder to see than open-cup nests, but this seems unlikely because cavity nests are often conspicuous against the uniform background of a tree trunk, and open nests are often built where they are hidden by vegetation. Finally, cavity nests are harder for olfactory predators to find.

There are several reasons why this third hypothesis might be true. First, some odorants produced by the incubating bird and the nest are absorbed by the walls of the cavity, and other odorants sink to the cavity floor, with the combined result that few odorants flow through the cavity hole to the outside air. Second, airflow out of a cavity with a small opening is irregular, and therefore the odor plume consists more of distinct puffs of odorants rather than a constant path. This makes it much harder for an olfactory predator to be able to track the odor plume to its source. Hence, both the first and third hypotheses probably explain why cavity nests suffer lower predation rates than open-cup nests.

Prediction 11.8

> In areas where there is a prevailing wind, especially at night, cavity nests should be oriented either directly toward it or away from it rather than at right angles to the wind. Cavity nests oriented directly toward or away from the wind will suffer lower predation rates from olfactory predators than cavity nests oriented at right angles to it.

Prediction 11.8 is based on the fact that when air flows around a tree trunk, the air immediately windward of the trunk is slowed by it and experiences turbulence (Figure 11.3). Likewise, directly leeward of the trunk, a wake zone occurs where the wind is slow, and turbulence predominates. Any cavity nests in these areas will experience intermittent gusts and low air velocities at their entrances. Hence, airflow out of the cavity is intermittent and does not create a steady odor trail that an olfactory predator can easily follow to its source. In contrast, air accelerates as it passes by the sides of a tree trunk, and this creates a vacuum that constantly draws air out of any cavities located on the trunk perpendicular to the wind (Figure 11.3). In this case, a strong and continual odor plume is created that an olfactory predator can easily follow to the nest. Hence, nest predation rates by olfactory predators should be lower on cavity nests on the windward and leeward sides of a tree trunk than those perpendicular to the wind. Henceforth, I refer to this as the *predator avoidance hypothesis*.

Many researchers have found that birds do not randomly locate their cavity nests on trees. The most widely accepted reason for this phenomenon is that birds orient their nests to create optimal temperatures for incubation (*temperature optimization hypothesis*), but studies that have tested this hypothesis have provided mixed results. In the hot deserts of Arizona, cavity entrances used by Gila woodpeckers (*Melanerpes uropygialis*) (Inouye et al. 1981) and elf owls (*Micrathene whitneyi*) (Hardy and Morrison 2001) are often located on the north side of cacti or trees, presumably because such cavities are cooler during the day than south-facing ones (Figure 11.4). These findings support the temperature optimization hypothesis, but in the same areas, both cactus wrens (*Campylorhynchus brunneicapillus*) and verdin (*Auriparus flaviceps*) build their nests facing southwest, where they are exposed to the hot afternoon sun (Austin 1974).

In cooler areas, sapsuckers (*Sphyrapicus* spp.) and flickers (*Colaptes* spp.) orient the entrances of their nest cavities in a southerly direction, presumably to maximize the amount of sunlight reaching them (Dennis 1971; Crockett and Hadow 1975; Inouye 1976; Wiebe 2001). These findings also support the temperature optimization hypothesis, but

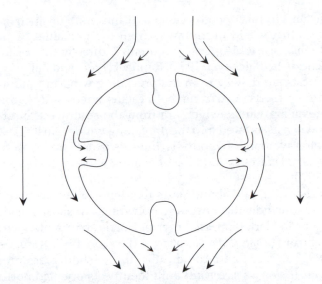

Figure 11.3 Airflow around a tree trunk that contains four cavities. Airflow out of cavities directly facing toward or away from the wind is slow and intermittent; airflow out of cavities that are perpendicular to the wind is fast and constant.

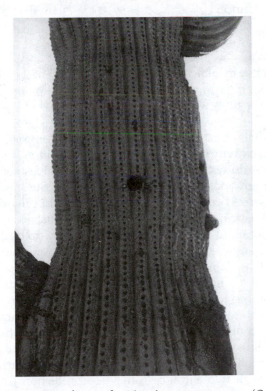

Figure 11.4 A nest cavity located on the north side of a saguaro cactus (*Camegiea gigantes*).

cavities also face south in Texas, where excess heat would presumably be a greater problem than the cold (Dennis 1971).

Unfortunately, wind direction has not been reported in most of these studies, so I cannot use them to test the predator avoidance hypothesis. Fortunately, there are some studies that do so. In Virginia, woodpeckers and flickers built their nests on the north-eastern side of trees that were directly away from the prevailing winds (Conner 1975). American kestrel (*Falco sparverius*) nests faced either directly into or away from the pre-vailing wind in Venezuela (Balgooyen 1990). Raphael (1985) and Balgooyen (1990) reported that woodpecker nests faced away from the prevailing wind in California. Austin (1974) noted that, in the hot deserts of Arizona, both cactus wrens and verdin build their nests facing into the prevailing winds, which are from the southwest. Nest success in cactus wrens was higher when nests faced into the prevailing winds (Austin 1974). These findings support the predator avoidance hypothesis, but this is not true of all findings. American kestrel nests in California face east, and the prevailing winds were from the south (Raphael 1985; Balgooyen 1990).

Several ground-nesting birds build nests that tend to face away from the prevailing wind; these species include the northern bobwhite, western meadowlark (*Sturnella neglecta*), eastern meadowlark (*Sturnella magna*), Harris sparrow (*Zonotrichia querula*), and white-crowned sparrow (*Zonotrichia leucophrys*) (Lanyon 1957; Rosenberry and Klimstra 1970; Norment 1993). But, not all birds do this. In the short grass prairies of Colorado, McCown's longspur (*Calcarius mccownii*) built their nests oriented north, and the nests of lark buntings face northeast. Yet, prevailing winds in the area were from the northwest or southeast (With and Webb 1993). In summary, the data indicate that many birds are specific about which side of a tree trunk should be used to excavate a cavity nest. In some species, that decision is based on thermoregulatory concerns, and in other species it is based on predator avoidance.

Prediction 11.9

> Predation rates by olfactory predators on cavity nests will be higher when cavities have more than one entrance, and therefore birds will avoid these cavities with multiple entrances.

We take it as a given that a nest cavity will have only one entrance, and I am unaware of any bird that excavates a cavity nest with two or more entrances. Furthermore, many birds that nest in cavities but do not excavate the nests themselves show a strong preference for cavities with one entrance and refuse to nest in cavities with multiple entrances (Hansell 2000). Yet, a second entrance to a cavity nest could provide an incubating bird a convenient means of escape if a predator appeared at the entrance to the cavity. Why then do cavity excavators not build a second entrance, and why do birds that build their nests in tree cavities that already exist avoid using cavities with more than one opening?

One hypothesis is that a benefit of a cavity nest is that it offers protection from olfactory predators because with a single entrance airflow out of the cavity is diminished and intermittent because air has to both enter and leave the cavity through the same opening. In contrast, air can easily flow through a cavity with multiple entrances, especially if the entrances are located on different sides of the tree. Hence, an olfactory predator would have a harder time following the odor plume emanating from a single-entrance cavity back to its source than following one coming from a multiple-entrance cavity.

To test if cavities with two entrances offer less protection from olfactory predators than those with one entrance, I ran an experiment using domestic dogs. I took circular flower pots (height = 0.15 m, diameter at the top = 0.17 m, and bottom diameter = 0.12 m) and cut a single hole (diameter = 0.04 m) halfway up the side of half of the pots. On the others, I cut two holes of the same size and height from the bottom. The two holes were on opposite sides of the pot. Before each test, I randomly selected one pot with a single hole and one pot with two holes and placed in each pot a dead starling that was provided by employees of the U.S. Department of Agriculture's Wildlife Services. I sealed the top of each pot with a wooden board so that air could leave or enter the pot only through the cut holes. The tests were conducted in pastures and fallow fields in Cache County, Utah, during 2004 through 2006. Seventeen dogs were used in these tests, and no dog was used more than once. Most, but not all, of the dogs had been trained to hunt birds.

To begin a test, I placed the two pots on the ground on an imaginary line perpendicular to the direction the wind was blowing and 10 to 20 m apart. After leaving the pots in place for 5 minutes, a dog was released by its owner 50 m directly downwind of the pots and allowed to search the area. I recorded the amount of time in seconds that elapsed before the dog found each pot. The test lasted 600 seconds, and if a dog failed to find a pot by the end of a test, then it was assigned a value of 600 seconds. I used a one-tailed, paired t test to assess if dogs found the two-hole pot sooner than the one-hole pot.

Dogs found the starling in the two-hole pot after a mean of 175 seconds (standard error = 51 seconds) and the one-hole pot after a mean of 260 seconds (standard error = 59). This difference was statistically significant ($t = 1.81$, $p = .04$). These results support the prediction that olfactory predators have a harder time locating cavities with only one hole than cavities with multiple openings.

Prediction 11.10

> Cavity nests with large entrances will suffer higher rates of nest predation by olfactory predators than those with small entrances. Therefore, birds will not build or nest in cavities that have large entrances.

Cavity excavators normally build entrances into which they can just barely squeeze, and cavities with large entrances are usually abandoned (Figure 11.5). Numerous studies have

Figure 11.5 Cavities with large entrances suffer higher rates of predation than those with small entrances and therefore are often abandoned by nesting birds. The bird in this cavity is a pileated woodpecker (*Dryocopus pileatus*). (Courtesy of the U.S. Fish and Wildlife Service's Bear River Migratory Bird Refuge.)

shown that cavity nests with large entrances suffer higher predation rates than those with small entrances (Caro 2005). One reason for this is because large entrances allow for more rapid airflow between the cavity and the outside air. Hence, olfactory predators will have an easier time locating and following an odorant plume from cavity nests with large entrances. However, another reason that cavity nests with large entrances suffer more predation is because it is less difficult for predators to gain access to their interior.

Prediction 11.11

> Anything that decreases the temperature differential between the inside of a cavity nest and the air immediately outside its entrance (especially at night) will decrease the vulnerability of that nest to olfactory predators. Two such variables are the temperature-insulating quality of the cavity walls and the volume of the cavity.

When the air is still, airflow between a cavity nest and the outside will be driven by temperature differentials between the outside air and air in the cavity nest. A simple mixing of the two air masses will not bring them into ambient condition because incubating birds and nestlings continually generate heat inside the cavity. How fast the cavity heats depends on the insulating quality of the cavity walls (live wood contains more water than dead wood and is a poorer insulator) and the volume of the cavity. The faster the temperature in the cavity rises, the more odorants will escape the cavity, and this will increase the risk of an olfactory predator finding the nest. Hence, predation rates by olfactory predators will be lower for nests built in large-volume cavities.

A few studies are germane to this last prediction. Nest predation rates on northern flickers (*Colaptes auratus*) declined as their nesting cavity became more voluminous (Wiebe and Swift 2001). Predation rates on the nests of pied flycatchers (*Ficedula hypoleuca*) and collared flycatchers (*Ficedula albicollis*) were higher (14%) in small nest boxes than in large ones (0%). Furthermore, flycatchers exhibited a strong preference for nesting in the larger nest boxes (87-cm^2 floor area) as 95% of them were occupied compared with an occupancy rate of 55 to 65% for smaller (57-cm^2) boxes (Gustafsson and Nilsson 1984).

Van Balen et al. (1982) showed that starlings prefer to nest in large cavities (>90 cm^3) than in smaller ones. In contrast, small cavities were more likely to be occupied by great tits than large cavities. This latter finding seems to contradict the hypothesis that birds should prefer larger cavities over smaller ones. However, Van Balen et al. also reported that great tits occupy small cavities because they are excluded from preferred cavities by larger birds with which they cannot compete. In the absence of competition, great tits preferred large cavities. Hence, these results support the predator avoidance hypothesis.

Prediction 11.12

> Nest predation rates by olfactory predators should be negatively correlated with the height of the nest from the ground.

It is more likely that the odorants from a high nest will remain above the olfactory zone of a predator and hence escape detection than a nest constructed close to the ground (Figure 11.6). Several studies have shown that cavity nests located close to the ground suffer higher predation rates than higher ones (Hansell 2000), especially from mammalian predators (Marzluff 1988; Caro 2005). One reason for this is that it is harder for terrestrial predators to reach high nests. Nevertheless, this is not the entire explanation because this pattern holds true even for nest predators that can climb (for example, raccoons and

Figure 11.6 The higher bird nests are built above the ground, the more likely their odor plume will be above the olfactory zone of mammalian predators.

opossums). For instance, predation rates on nests of Baya weavers decreased with their height above the ground even though most nests were depredated by snakes and rodents, both of which can easily climb trees and reach the nests (Quader 2006).

There is a price to be paid, however, for placing nests high above the ground. Such nests are more vulnerable to high winds (Marzluff 1988; Caro 2005) and to avian predators (Yahner and Cypher 1987; Caro 2005).

Prediction 11.13

> Nest cover, which is a measure of how much of the nest is obscured by obstacles (usually vegetation) in the immediate vicinity (1 m) of the nest, will influence the probability of nest depredation by visual predators but not olfactory predators.

Evidence for this prediction is strong. Martin (1992) conducted a literature search and found that 29 of 36 studies reported that predation rates were lower at nests with greater concealment. Clark and Nudds (1991) evaluated 38 nesting studies and concluded that nest concealment reduces predation, but its effectiveness is dependent on which predator species were depredating nests. Concealment was particularly important when bird predation was prevalent but less important when mammals were present (Clark and Nudds 1991). Schranck (1972) found that nest concealment was unimportant for both artificial and natural duck nests in North Dakota where mammalian predators dominated. Keith (1961) also noted that, in Alberta, where most nest predation was by skunks (Figure 11.7), depredated duck nests were as well concealed as successful nests. Concealment of field

Figure 11.7 Skunks hunt at night and use olfaction to locate bird nests.

sparrow nests did not influence their probability of depredation by snakes (Best 1978). Bowman and Harris (1980) discovered that concealment of the nest did not influence nest predation by raccoons except when concealed nests were compared to nests that were completely in the open. The authors attributed this to the fact that raccoons were locating nests by olfaction, and that nests that were hidden from view were still easy for them to find. Hence, most of these results support the prediction.

chapter twelve

Do weather, convection, isolated surface features, or shelterbelts influence nest predation rates of olfactory predators?

Impact of weather on olfactory predators

Prediction 12.1

> Nest predation rates by olfactory predators should be lower where and when the wind is still at night.

When the air is completely still, odor plumes emanating from nests will be circular and short. This will reduce the probability of an olfactory predator intercepting the odor plume while hunting compared to when the odor plume is long and narrow. Furthermore, an olfactory predator will have a harder time finding nests when the air is still because the predator cannot locate the nest by simply moving upwind. Unfortunately, no studies have compared the probability of a nest surviving a night when there is no wind to when there is a breeze, but this would be easy to test using captive predators.

Prediction 12.2

> When there is a breeze at night, nest predation rates by olfactory predators will be lower when and where the wind is strong and gusty.

In areas that normally are windy at night, nest predation rates by olfactory predators should be lowest in those sites where wind velocities exceed 20 km/hour and where turbulence is maximized. These areas would include on hilltops, above cliffs, or along draws where mountain breezes occur. I am unaware of any studies that have tested this prediction.

Prediction 12.3

> When and where inversions occur at night, nests located above them will experience lower predation rates from olfactory predators than nests located below the inversions.

When a low-lying inversion hovers low to the ground, a nest located above it in a tree or on a hill should be undetectable to olfactory predators hunting beneath the inversion. Unfortunately, no one has examined whether nests located above inversions are safer at night than nests located beneath the inversion.

Prediction 12.4

> Water will displace odorants from binding sites on the surface of animals, causing the release of odorants into the atmosphere. Hence, nest predation rates by olfactory predators should be higher when chicks or incubating birds are wet. This is likely to happen when it is drizzling, foggy, or humid.

There is good evidence that supports this prediction. Search dogs trained to locate a missing person or a cadaver (Figure 12.1) can do so faster when barometric pressure is low (Shivik 2002); low barometric pressure is correlated with humid conditions. Several authors have observed that bird dogs, foxhounds, and trailing dogs can locate their quarry best when it is cloudy and humid but have difficulty when it is hot and dry (Budgett 1933; Johnson 1977; Lowe 1981).

Figure 12.1 Search dogs are used to find a missing person. (Courtesy of John Shivik.)

Nest predation rates also increase during wet weather. Palmer et al. (1993) report that in Mississippi rainfall increased the probability that a turkey nest would be depredated. Similarly, in New York State, where nest predation was the major reason why turkey nests failed to hatch, annual variation in the survival rates of turkey nests was positively correlated with the annual amount of rainfall during the incubation period (Roberts et al. 1995; Roberts and Porter 1998a, 1998b).

Roberts et al. (1995) and Roberts and Porter (1998b) hypothesized that moisture increased the efficiency of nest predators, but they were unsure why this relationship was true. They speculated that bacteria on the bird's surface or on skin cell rafts grow faster during humid conditions, and therefore produce more odorants. I believe that a more likely explanation is that a wet bird releases more odorants than a dry one because water competes with odorants for binding sites on a bird's feathers and skin. Hence, rainfall increases the odorants released by the incubating bird and creates a stronger odor plume. Support for my hypothesis comes from the easily observed fact that humans cannot detect much odor from a dry animal. However, if an animal is saturated with water, its odor becomes easily detectable to humans. If you want to verify this for yourself, then give your pet a bath. Its odor will become more pronounced a few seconds after it has gotten wet, yet bacterial populations would not increase over such a short period.

Impact of convection on olfactory predators

Prediction 12.5

> Over-water nests should suffer lower predation rates by olfactory
> predators than upland nests.

Most olfactory predators hunt at night when air over warmer bodies of water is drawn to the center of the water body and then rises into the air (i.e., a land breeze). Hence, olfactory predators along the shoreline would be unable to smell an over-water nest. Therefore, over-water nests should experience lower predation rates from olfactory predators than upland nests (Figure 12.2). In fact, this prediction is accurate. Predation rates on over-water nests or nests located in swamps and marshes are often lower than those located in drier sites (Janzen 1978; Picman 1988; Arnold et al. 1993; Maxson and Riggs 1996; Honza et al. 1998). However, the difference in predation rates between wet and dry sites is usually attributed to the water serving as a barrier to mammals, making it hard for them to reach over-water nests (Lokemoen and Woodward 1993; Lokemoen and Messmer 1994; Maxson and Riggs 1996).

Both factors (nocturnal updrafts and water barriers) should help protect over-water nests, but the relative importance of each is unknown. However, the relative importance of these two factors can be teased apart. If water serves primarily as a barrier to predators, then nest predation rates should remain high when the water is shallow enough for predators to wade to them or when nests are close to shore. If nocturnal updrafts are important, then water depth and distance from shore should be unimportant. Experiments have yielded mixed results. Giroux (1981) found that duck nesting success on islands was related to the islands' distance from shore but not water depth. Krasowski and Nudds (1986), Maxson and Riggs (1996), and Brua (1999) reported that nest success of ducks was unrelated to either water depth or distance to shore, but Albrecht et al. (2006) found the opposite.

Another way to tease apart the relative importance of land breezes in protecting over-water nests is to compare nest predation rates when and where land breezes occur to

Figure 12.2 Water in this beaver pond is warmer than the adjacent ground. This causes mist to occur over the pond as the warmer air rises and cools. This causes a local land breeze and updraft over the pond. Hence, predators standing on the shoreline will have difficulty detecting any odor from the wood duck located in the nest box. (Courtesy of Gary San Julian.)

times and places where they are absent. Land breezes will occur only when the water temperature is warmer at night than the land surface and when atmospheric conditions allow for their formation. Hence, it should be easy to compare predation rates of over-water nests during nights when there is a land breeze and when there is not. Unfortunately, this comparison has yet to be made.

Prediction 12.6

> Nests located on the south side of a hill are less likely to be depre-
> dated by olfactory predators than nests located on the north side of
> the same hill. Therefore, nest densities should be higher in the former
> areas.

Komar (1999) and Conover (unpublished manuscript, 2006) noted that the ability of dogs to detect odor plumes decreased when temperatures were above 29°C. If the olfactory ability of other mammalian predators also is affected by heat, then south-facing slopes would be safer places to hide from them than north-facing slopes because south-facing slopes will be hotter when it is sunny (Figure 12.3). Furthermore, south-facing slopes will produce more and stronger updrafts than north-facing slopes, and these updrafts are likely to cause odor plumes to rise above the olfactory zone of mammalian predators.

At night, south-facing slopes also should retain heat longer than north-facing ones. Hence, updrafts should continue to occur on south-facing slopes, at least during the early evening. For these reasons, nests located on south-facing slopes should suffer lower predation rates from olfactory predators than those on north-facing slopes.

This prediction is an interesting one to test because north-facing slopes are more thickly vegetated, at least in arid regions, than south-facing slopes. Hence, birds should nest on north-facing slopes to hide from visual predators and nest on south-facing slopes to hide from olfactory predators. In Chapter 5, I described an experiment in which my

Figure 12.3 South-facing slopes are exposed to more sunlight and are warmer than north-facing slopes. (Courtesy of Madsen.)

results demonstrated that predation rates on artificial nests located on south-facing slopes were lower than those on north-facing slopes.

On sunny days, bare rock or soil surfaces will be hotter than vegetated surfaces and should experience more updrafts that may draw the odor of nearby nests above the olfactory zone of predators. Perhaps this is why successful bobwhite nests in Texas were more likely to be near bare ground than unsuccessful nests (Lusk et al. 2006). The authors noted that this difference might be related to different thermal patterns between nests by bare ground and those by vegetated surfaces but did not speculate regarding why thermal differences would be important.

Impact of isolated surface features on olfactory predators

Prediction 12.7

> Olfactory predators should have a harder time locating nests on hills, especially nests on hilltops. Therefore, predation rates by olfactory predators should be lower, and nest densities should be higher on hills than on level ground.

Hills create turbulence, especially when airflow is fast enough to cause boundary layer separation, and birds should take advantage of this by hiding and locating their nests there (Figure 12.4). Nests on hills are also more likely to be situated above any low-lying inversion. There are a few studies that can be used to test this prediction. Tremblay et al. (1997) found that in Canada where the arctic fox (*Alopex lagopus*) was the primary predator, nests of greater snow geese (*Chen caerulescens atlantica*) located on hillsides were more successful than those located in lowlands. Skeel (1983) reported that most whimbrels (*Numenius phaeopus*) nesting in hummock-bog and sedge-meadow habitats of Manitoba

Figure 12.4 Turbulence and wind velocities are higher on the tops of hills than in flat terrain. (Courtesy of Madsen.)

locate their nests on prominent surface features (that is, hummocks and ridges) that protruded above the surrounding area. Interestingly, these nests were not located at random on these surface features. Instead, they tended to be on the leeward side of the surface features where boundary-layer separation should occur (Chapter 6).

> *Prediction 12.8*
>
> > Ground-nesting birds should locate their nests near isolated surface features, and those birds that do so should experience lower predation rates by olfactory predators than birds that locate their nests in the open.

Isolated surface features that are impermeable to the wind create turbulence in their vicinity. Therefore, olfactory predators should have a more difficult time locating a nest there than in an open area (Figure 12.5). Nests of mountain plovers (*Charadrius montanus*) and other Charadriidae are often located near piles of dung from cattle or other animals (Graul 1975). McCown's longspur (*Calcarius mccownii*) and horned larks (*Eremophila alpestris*) not only locate their nests near piles of cattle dung but also nest close to other surface features, such as rocks and cacti (With and Webb 1993; With 1994). However, longspur nests near cacti suffer higher predation rates than those in the open (With 1994). In Texas, northern bobwhites (*Colinus virginianus*) prefer to locate their nests near the base of prickly pear cactus (*Opuntia* spp.), but these nests are no more successful than nests located elsewhere (Figure 12.6) (Hernandez et al. 2003). Evening grosbeaks (*Coccothraustes vespertinus*) nests located close to a tree trunk are more successful than those located farther away (Bekoff et al. 1989). Blackstarts (*Cercomela melanura*) build substantial stone ramparts around their nests (Leader and Yom-Tov 1998). The ramparts have the effect of creating turbulence and slowing any predators that try to break into the nest. Ruffed grouse (*Bonasa umbellus*) nest on the ground and prefer to site their nests adjacent to large trees or stumps or fallen logs, and grouse that do so are more successful than those nesting elsewhere (Bump et al. 1947; Redmond et al. 1982; Bergerud and Gratson 1988).

Figure 12.5 Isolated surface features such as this boulder create local turbulence. (Courtesy of Madsen.)

Figure 12.6 In Texas, quail often build nests at the base of prickly pear cactus.

Prediction 12.9

> An isolated tree, bush, or other surface feature should create a local turbulence. Therefore, predation rates by olfactory predators should be lower on nests located under trees and bushes, but this will be offset by higher predation rates by visual predators, which are likely to use the surface features for perch sites when hunting.

Many avian species prefer to build nests under trees and bushes (Caro 2005), but less is known how predation rates on these nests compare to those located away from trees or bushes. Whimbrels locate their nests near shrubs (Skeel 1983), as do McCown's longspur

Figure 12.7 Coyotes are more likely to hunt near brush than out in the open. (Courtesy of Madsen.)

(With 1994). However, ground nests of McCown's longspur suffered a higher nest predation rate when nests were located within 1 m of shrubs than when placed out in the open
because the main nest predator, the 13-lined ground squirrel (*Spermophilus tridecemlinea-
tus*), concentrates its foraging close to bushes (With 1994). Coyotes may do likewise.
Guthery et al. (1984) reported that coyotes were also more likely to visit meat bait stations
if they were located near brush rather than in the open (Figure 12.7). These findings do
not support this prediction.

Impact of shelterbelts on olfactory predators

Prediction 12.10

> Olfactory predators will have difficulty using odor plumes to locate
> prey in the region extending from $-2h$ windward of a shelterbelt to
> $20h$ leeward of it. Shelterbelts, however, will attract avian predators.
> Hence, nests near shelterbelts will experience lower predation rates
> from olfactory predators but higher predation rates from visual
> predators than nests located in the open.

A good place to hide from olfactory predators is on the leeward side of the shelterbelt
because this is where updrafts will predominate. However, shelterbelts should also attract
avian predators because they provide the perch sites that avian predators need for hunting
(Figure 12.8). Hence, shelterbelts help protect nesting birds from one type of predator
while enhancing their vulnerability to another. Perhaps the best place to nest near a
shelterbelt, given this circumstance, is in the wake zone, which is 10 to $20h$ downwind of
the shelterbelt. This region is still close enough to the shelterbelt to experience turbulence
but may be far enough away for an avian predator perched in the shelterbelt to see the
nest or incubating bird. Few studies have compared nest predation rates as a function of
their distance from a shelterbelt or identified the relative risk posed by olfactory or visual
predators. The one study on this topic of which I am aware showed that bobolinks

Figure 12.8 Most avian predators hunt from perches. Hence, birds nesting near windbreaks may be vulnerable to them. (Courtesy of U.S. Fish and Wildlife Service's Bear River Refuge.)

(*Dolichonyx oryzivorus*) avoided nesting close to some shelterbelts but not others (Bollinger and Gavin 2004). The same study found that nests within 50 m of a shelterbelt had a lower survival rate than those located more than 100 m from it.

> *Prediction 12.11*
>
> Predation rates by olfactory predators should be lower near shelterbelts with dense canopies rather than ones with porous canopies.

Olfactory predators should have an easier time finding nests around porous shelterbelts than dense ones because turbulence is positively correlated with shelterbelt density. Shelterbelt porosity should not influence nest predation rates by visual predators. Hence, nest predation rates should also be lower near shelterbelts with dense canopies. I am unaware of any studies that have tested this prediction.

> *Prediction 12.12*
>
> Predation rates by olfactory predators should be lower where shelterbelts are perpendicular to the prevailing nocturnal wind rather than parallel to it.

Olfactory predators will have a harder time finding prey around shelterbelts built so that they are perpendicular to the prevailing nocturnal winds because shelterbelts so oriented will create more turbulence and updrafts than those that are parallel to the wind. This may initially seem a difficult prediction to test because shelterbelts are usually built to block the wind and are built perpendicular to it. Yet, there are many shelterbelts that were not really built to block the wind. Instead, they came into being because many agricultural fields are bordered by a thin corridor of brush and trees, and many old fences have become overgrown with similar vegetation. Some of these volunteer shelterbelts will be parallel to the prevailing winds. Hence, this prediction will not be difficult to test, but it has not been tested to date.

Prediction 12.13

> The ability of shelterbelts to hinder the ability of olfactory predators
> to depredate nests will be correlated with wind velocity.

This prediction is similar to Prediction 12.12 except now the comparison is across time rather than across shelterbelts. This also is an easy prediction to test because there will be some nights when the wind velocities are high and other nights when the air is still. By comparing predation rates during nights with these various wind conditions, it should be possible to test this prediction.

Prediction 12.14

> The width of a shelterbelt, the plant species within it, or its other
> characteristics will not have an impact on the nest predation rates
> of olfactory predators unless those characteristics change the shel-
> terbelt's height or porosity.

The only characteristics that influence the amount of updrafts and turbulence caused by a shelterbelt are the height of its canopy and its porosity (Figure 12.9). Hence, these should be the only characteristics that will have an impact on the ability of olfactory predators to locate nests. As far as I am aware, this prediction has not been tested.

Figure 12.9 It should not matter if a shelterbelt is composed of a single plant species, as is true of this shelterbelt, or of multiple plant species.

chapter thirteen

Do prairies, savannas, forests, or edge habitats influence nest predation rates of olfactory predators?

Nest predation by olfactory predators in prairies and open fields

Prediction 13.1

> If a nest is surrounded by vegetation that can impede the movements of mammalian predators, then the vegetation's effectiveness in reducing nest predation by olfactory predators will depend on how far it extends from the nest.

Some vegetative cover may be of such density that the open spaces are not large enough for a midsize mammalian predator to slip through and of such stiffness that a predator could not easily bend the plants out of the way. In such cases, a foraging mammal may avoid these areas because the energetic cost of moving through them may exceed the potential gain of searching them for food. I have seen some stands of phragmites that were thick enough to discourage raccoons from searching them thoroughly for food. Likewise, nests surrounded by thorny vegetation, such as blackberries or cacti, may gain some protection from these plants because of the pain they can inflict on a passing predator (Figure 13.1). For instance, red-legged partridge (*Alectoris rufa*) prefer to nest in stands of stinging nettle (*Urtica dioica*), but it is not known if nettle deters nest predators (Rands 1986). In the tropics, some canopy-nesting birds (for example, the Baya weaver) prefer to nest in trees that have thorns on their trunks and branches. Birds that nest in these trees suffer less nest predation than those that nest in other trees (Quader 2006).

Once an olfactory predator has detected an odor plume from an incubating bird, it will probably endure much discomfort and pain to reach the nest and obtain food. Hence, briar patches may only be effective against a foraging predator that has not detected a nest's odor plume. The closer a predator can get to a nesting bird without having to enter the briar patch, the more likely it is to detect its odor plume. Hence, wide patches of briars or cacti should provide more protection from olfactory predators than narrow barriers (Figure 13.2). Although this prediction seems reasonable, I am unaware of any studies that have tested it.

Figure 13.1 Foxes and other midsize mammals have little difficult moving through dense cover. (Photo courtesy of Jaimie Jimenez.)

Figure 13.2 A wide patch of cactus helps protect nesting birds from midsize predators, provided that a predator standing outside of it cannot detect the odor plume of a bird nesting within it.

Prediction 13.2

Nest predation rates by olfactory predators and nest densities will be correlated with the surface roughness (d and z_0) of the habitat patch where the nest is located. Predation rates by visual predators will be less affected by changes in d and z_0.

Figure 13.3 This dense cover of tall grass was created specifically to protect upland-nesting ducks from predators. (Courtesy of the U.S. Fish and Wildlife Service's Bear River Migratory Bird Refuge.)

Prediction 13.3

> Dense cover in the habitat patch surrounding nests reduces preda-
> tion rates by olfactory predators but does so not by impeding the
> movements of predators but by increasing turbulence and updrafts.

Many wildlife managers believe that dense nesting cover reduces nest predation rates (McKinnon and Duncan 1999; Jimenez and Conover 2001). In fact, Ducks Unlimited, the U.S. Fish and Wildlife Service, state wildlife agencies, and many landowners try to improve the nesting success of upland-nesting ducks by spending their time and money to create and maintain patches of dense nesting cover (Figure 13.3). Yet, many studies have reported no relationship between dense nest cover and breeding success in upland-nesting ducks, especially in areas where olfactory predators pose a greater risk than visual predators (Byers 1974; Jimenez and Conover 2001).

Many authors have hypothesized that dense cover protects nests from mammalian predators by physically impairing the movements of skunks, raccoons, and foxes (Dueb-bert 1969; Schranck 1972; Livezey 1981; Lariviere and Messier 1998). Yet, observations of foraging predators indicate that they have little problem of moving through dense cover (Bowman and Harris 1980; Crabtree et al. 1989). In fact, raccoons searched areas with dense cover more intensely than areas with light or no cover (Bowman and Harris 1980). Furthermore, Jimenez (1999) measured the extent to which vegetation of different habitat patches impeded the movements of a meso-predator and found that resistance to predator movements was unrelated to duck nesting success.

Other authors have assumed that dense nesting cover protects nests by increasing structural heterogeneity. Heterogeneity in turn is believed to increase the number of potential nest sites a predator must check and thereby increases the amount of time it must spend searching to find a single nest (Mankin and Warner 1992). However, olfactory predators do not have to check each potential nest site individually; rather, they can search for nesting birds at a distance by seeking odor plumes.

Figure 13.4 The roughness of forest ground surfaces increases when covered by stumps, logs, and downed trees. (Courtesy of Lee Stribling.)

A third hypothesis suggested by the olfactory concealment theory is that dense cover in the habitat patch functions by increasing the habitat's d and z_0. Doing so would increase the frequency of updrafts and turbulence and make it harder for olfactory predators to find nests. Although few ornithological studies have measured the d and z_0 of habitat patches, several have examined the fate of nests located in rough habitat compared to those in other areas. For instance, Tremblay et al. (1997) determined that snow geese (*Chen caerulescens*) nesting in the Northwest Territories of Canada suffered less predation when the surface roughness within 30 m of their nest was high. In their study site, variance in surface roughness resulted from an irregular ground surface created by frost heaves, polygon ridges, and rocks.

Tirpak et al. (2006) measured the density of stumps and fallen logs in forest patches and compared ruffed grouse nest densities and nest success in patches in stands with high densities of coarse woody debris to stands with low densities. They found that both grouse nest densities and nest success were higher in stands with high amounts of coarse woody debris on the ground (Figure 13.4). Bowman and Harris (1980) showed, by creating large piles of tree branches and planting rows of mature saw palmettos (*Serenoa repens*) within pine and oak forests, that the ability of raccoons to locate artificial quail nests decreased as the forest floor became rougher.

Prediction 13.4

> Olfactory predators will have an easier time locating nests on prairies that are flat than on prairies that are hilly, have a broken terrain, or vary in topography.

Some prairies, such as large parts of the Great Plains, are generally flat; other prairies, like the Nebraska Sand Hills and Washington's Palouse Prairie, consist of large expanses of hills, ridges, or ancient sand dunes. A flat prairie has a lower z_0 than one that is hilly because the hills and ridges provide friction to airflow. Hence, airflow will be less turbulent over flat terrain; therefore, nest predation rates by olfactory predators should be higher there than in areas with a broken or hilly terrain. This prediction remains to be tested.

Prediction 13.5

> Grass-dominated patches are more likely to produce updrafts than forb-dominated patches. Therefore, nests in grass-dominated areas will experience lower rates of predation by olfactory predators than nests in forb-dominated areas.

Grass leaves tend to deflect airflow upward because of the way they bend. Hence, grass patches are more likely to produce updrafts than patches dominated by other plants, such as forbs. In Illinois, Best (1978) found that mammalian predation on nests in grass was lower than on nests located in areas dominated by forbs or shrubs. Snake predation, however, was similar across both habitats. In Iowa, field sparrow (*Spizella pusilla*) nests located in grassy areas suffered lower predation rates by midsize mammals (for example, skunks and raccoons), small mammals, birds, and snakes than in areas with forbs (Best and Stauffer 1980). In the mixed-grass prairies of North Dakota, predation rates on the nests of clay-colored sparrows (*Spizella pallida*) and vesper sparrows (*Pooecetes gramineus*) declined with increasing amounts of Kentucky bluegrass (*Poa pratensis*) within 10 m of their nest (Grant et al. 2006). In England, gray partridges (*Perdix perdix*) preferred to nest in areas where there was an abundance of dead grass. The quantity of dead grass was an important factor in predicting the likelihood of nest depredation by a predator (Rands 1982). However, grass cover does not always provide better protection from predators than forbs. In tall grass prairies, nest success of clay-colored sparrows, savannah sparrows (*Passerculus sandwichensis*), and bobolinks was not related to the proportion of grass in a habitat patch (Winter et al. 2005).

Prediction 13.6

> Updrafts will be more pronounced in fields dominated by bunch grass than in fields dominated by turf grass. Therefore, nest predation rates by olfactory predators will be lower in fields of bunch grass.

The leaves and stems of bunch grass radiate out from a central clump. Such clumps often are spaced apart, and these two growth characteristics create a rougher canopy than sod-forming grass species, such as Kentucky bluegrass. Hence, both d and z_0 are higher in fields dominated by bunch grass than turf grass, and because of this, olfactory predators should have a harder time locating nests in fields of bunch grass (Figure 13.5).

Best (1978) reported that, at his field site in Illinois, mammalian predation of field sparrow nests was rare in areas surrounded by Indian grass (*Sorghastrum nutans*), which is a tall bunch grass. Mammalian predation was higher on nests located in Kentucky bluegrass. Nesting success of lesser prairie chickens (*Tympanuchus pallidicinctus*) was higher when hens selected sand bluestem (*Andropogon hallii*) for nest cover rather than

Figure 13.5 Bunchgrass and other grass that grows in clumps creates a rough surface (note the nest at the base of the grass clump).

other grass species (Riley et al. 1992). The authors described sand bluestem as forming tall, widely spaced clumps with spreading stems. Sveum et al. (1998) reported that greater sage-grouse exhibited a nesting preference for areas with bunchgrass over areas with other grasses, such as Sandberg's bluegrass (*Poa sanbergii*) and cheatgrass (*Bromus tectorum*). Nest success was highest in a big sagebrush/bunchgrass habitat type. Northern bobwhites preferred to nest in areas with bunchgrass, such as little bluestem (*Schizachyrium scoparium*) and grama grass (*Bouteloua* spp.), and nests located there were more successful than nests located elsewhere (Slater et al. 2001; Hernandez et al. 2003). Dickcissel (*Spiza americana*) nests built in grass tussocks of *Setaria faberi* and *Polygonum pennsylvanicum* were more successful and suffered less predation than those built in forbs (Harmeson 1974).

Prediction 13.7

> Habitats dominated by tall grasses will reduce nest predation by midsize olfactory predators, such as raccoons; areas dominated by short grasses will not.

The height of an updraft produced by a grassy habitat is correlated with the height of the canopy. If a predator's olfactory zone is lower than the height of the grass, then updrafts produced by the grass should reduce the predator's ability to detect and locate a nest in that habitat. Hence, even a short grass, such as colonial bentgrass (*Agrostis tenuis*) or

Figure 13.6 Grass-dominated habitats will be more effective in protecting ground-nesting birds from predators when the cover is above the predator's head.

Kentucky bluegrass, should be able to elevate the scent of a nest above the olfactory zone of a snake or rodent. But, it would require a tall grass with leaves close to a meter in length to produce updrafts of sufficient height to elevate odorants from a nesting bird above the olfactory zone of a raccoon or fox (Figure 13.6).

Numerous studies in the short grass and tall grass prairies of North America have shown that nest success of ground-nesting birds increases with grass height (Taylor et al. 1999; Howard et al. 2001; Jimenez and Conover 2001; Holloran et al. 2005; Manzer and Hannon 2005; Lusk et al. 2006). Tall grass could protect nests through two different mechanisms: (1) Tall grass at the nest site would help conceal the nest from visual predators, and (2) tall grass located throughout the habitat patch would increase updrafts and help protect the nest from olfactory predators. At least for greater sage-grouse, the second mechanism is more important. Sveum et al. (1998) and Aldridge and Brigham (2002) report grass height within 1 m of the nest did not influence the nest success of sage-grouse, but grass height in the nesting habitat patch did (Figure 13.7).

Figure 13.7 Greater sage-grouse. (Courtesy of Justin Harrington.)

In England, the nesting success of black grouse (*Tetrao tetrix*) was higher in pastures where the mean height of the vegetation was 34 cm rather than in pastures where the vegetation height was only 22 cm (Calladine et al. 2002). Likewise, nest success was higher in ungrazed than grazed moors where the grass was 32% shorter (Baines et al. 1996).

Prediction 13.8

> Nest predation rates by olfactory predators should be less on nests located along the edges of an open field than in the interior of the patch.

Turbulence and updrafts will occur along ecotones wherever the two adjacent habitats differ in surface roughness (Chapter 7). Johnson (1977) reported that dogs have difficulty following scent when they cross from one habitat type to another, such as from a plowed field to a grassy pasture or from a field of wheat to one of oats or rye. He observed that, along such ecotones, dogs may either lose the scent entirely or take longer to find it (Johnson 1977). However, in the prairies of the Great Plains, predation rates on artificial nests (Pasitschniak-Arts and Messier 1995), duck nests (Jimenez 1999), and greater prairie chicken nests (McKee et al. 1998) either increased or did not vary with distance to an ecotone.

Prediction 13.9

> Predation rates by olfactory predators will be lower on open field nests that are near a wood lot than on open field nests that are far away from wood lots.

When air flows from an open field to a forest, the air will start to rise before it reaches the forest. This region of open habitat next to a forest where updrafts occur (a distance up to five times the height of the forest canopy) should afford nests more protection from olfactory predators than other parts of the same open field. Concomitantly, the trees along the forest's edge will provide good perching sites for predatory birds. Hence, this same region will provide less protection from visual predators than other parts of the field. The relative safety of nests located in the middle of open fields versus those near forests should depend on the relative risk posed by olfactory versus visual predators.

Some authors report that predation rates on passerine nests located in grasslands increase with decreasing distance from a wooded edge (Gates and Gyusel 1978; Johnson and Temple 1986, 1990; Bollinger and Gavin 2004) or a shrubby one (Winter et al. 2000); others report the opposite (Grant et al. 2006). Yet, in a meta-analysis of the scientific literature, Lahti (2001) concluded that nest predation in grasslands did not differ between nests near forest edges and those away from forest edges. However, there is no evidence that grassland nests experience less predation as proximity to a forest edge increases. Hence, this prediction is not supported.

This lack of an effect may be because protection from olfactory predators resulting from updrafts near a forest edge is negated by greater activity of avian predators along woody and shrubby edges (Angelstam 1986; Andren and Angelstam 1988; Andren 1992; Paton 1994; Soderstrom et al. 1998) as well as higher densities of small mammalian predators, such as red squirrels (*Tamiasciurus hudsonicus*) and eastern chipmunks (*Tamias striatus*) along forest edges (King et al. 1999).

Nest predation by olfactory predators in savannas

Prediction 13.10

> The vulnerability of savanna nests to olfactory predators should be positively correlated with the savanna's z_0, and the savanna's z_0 should be highest when the proportion of brush in the habitat is neither too thick nor too thin.

Langen et al. (1991) compared predation rates on artificial bird nests placed in chaparral that varied in brush cover from 100 to 5%. They found that predation rates were lowest in habitat patches where brush levels were less than 25%. In their study site, most nests were depredated by mammals, mostly rodents, opossums, and raccoons. In mixed-grass prairies, predation rates on nests of vesper and clay-colored sparrows declined as the percentage of cover in tall shrubs increased from 0 to 30% (Grant et al. 2006).

Mallards, northern pintails (*Anas acuta*), and gadwalls (*Anas strepera*) prefer to nest in shrub savannas rather than open grasslands (Duebbert et al. 1983; Greenwood et al. 1995; Shaffer et al. 2006). Predation rates on nesting ducks and geese are often lower in open areas with brush than in open areas without brush (Cowardin et al. 1985; Sugden and Beyersbergen 1987; Jackson et al. 1988; Tremblay et al. 1997).

Sharp-tailed grouse preferred to nest in habitat patches with some trees rather than in open patches. Their nest success also was positively correlated with tree density within 0.5 km of the nest, although perch sites close to the nest made them more vulnerable to predation by corvids (Manzer and Hannon 2005). Nest success rates and nest densities of whimbrels were higher in areas where there were scattered brush, stunted trees, and patches of large hummocks up to 75 cm high than in sedge-meadows, flat and treeless areas, or heath-tundra habitats (Skeel 1983). In contrast, the habitat around successful nests of northern bobwhite had less shrub cover (0.3%) than the habitat around unsuccessful nests (4.5% shrub cover) (Taylor et al. 1999).

Nest predation by olfactory predators within forests

Prediction 13.11

> Ground and shrub nests located in areas where the upper surface of the forest canopy is rough are less likely to be depredated by olfactory predators than nests located in areas where the upper surface of the forest canopy is smooth.

Baldocchi and Meyers (1988) observed that tree crowns that protrude above the forest canopy cause local wakes, sweeps, and vortices. Hence, wind blowing past these protruding trees is more likely to be directed downward into the forest canopy and the subcanopy space. In such areas, sweeps and ejections will increase turbulence and updrafts along the ground, causing odor plumes to rise and disperse. For these reasons, nest predation rates by olfactory predators should be less in forests where the upper surface of the forest canopy is rough. Therefore, ground-nesting and shrub-nesting birds should prefer to nest in these areas. It is not intuitive that the texture of the upper canopy surface should affect predation rates on nests located many meters below it. Hence, it is not surprising that this prediction has never been tested. Yet, it is possible to make some specific predictions about where and when rough canopy surfaces should occur. Therefore, I can make additional predictions about where birds should nest based on the texture of the upper canopy surface and then determine if the scientific literature supports the predictions. These predictions are listed next.

Prediction 13.12

> Naturally regenerating forests will experience more turbulence and updrafts than an even-aged forest plantation. Hence, nest predation rates by olfactory predators should be lower on ground and shrub nests located within naturally regenerated forests than within even-aged forests.

Northern bobwhite avoided nesting in even-aged, closed-canopy forests of loblolly pine (White et al. 2005). Instead, they preferred to nest in open-canopy pine forests. Black grouse and capercaillie (*Tetrao urogallus*) avoided nesting in even-aged plantations of Norway spruce and Scotch pine, preferring instead multiaged conifer forests (Storaas and Wegge 1987). Nest losses, however, were similar in all habitats.

DeGraaf (1995) studied predation rates on artificial ground and shrub nests in the northern hardwood forests of New Hampshire and found no differences in nests located inside even-aged forest stands created by clearcuts and nests in multiaged stands created through natural regeneration. In his study area, nests were depredated by mammals, mostly large ones such as raccoons, black bears (*Ursus americanus*), and fishers (*Martes pennanti*).

Prediction 13.13

> Nest predation rates by olfactory predators should be lower for ground and shrub nests in even-aged forests after rather than before the trees have been thinned. This effect will be correlated with severity of the thinning.

The more severely a tree stand is thinned, the more sweeps and ejections will occur with the subcanopy space and along the ground. This in turn will make it harder for olfactory predators to locate nests. There is some support for this prediction. Pierre et al. (2001) distributed artificial nests located inside nest boxes in an uncut forest and a severely thinned forest where 92% of the trees had been removed. They discovered that predation rates by mammalian predators were several times higher in the unthinned forest than in the thinned areas (Figure 13.8).

Prediction 13.14

> Nest predation rates by olfactory predators should be lower along the edge created where two forests of different heights meet.

Research does not support this prediction. Gibbs (1991) reports that in Costa Rica nest predation on artificial nests was lower in a mature forest than along the interface of the mature forest and a second-growth forest less than ten years old. The author believes that this resulted because predators from the mature forest often traveled to the second-growth

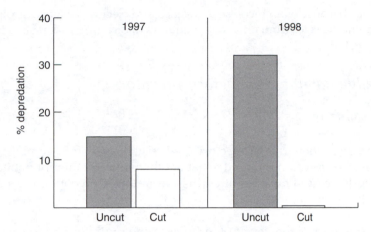

Figure 13.8 Predation rates on artificial nests in an uncut forest and a heavily thinned forest where 92% of the trees had been removed. (Based on data from Pierre, J.P., H. Bears, and C.A. Paskowski, *Auk* 118:224–230, 2001.)

forest where food was more abundant. Lahti (2001) reports that five studies conducted elsewhere found no increase in nest predation near edges where forests of two different heights met, but these studies also found no decrease in nest predation.

> *Prediction 13.15*
>
> Nest predation rates by olfactory predators should be lower in diseased forests where canopy trees are dead or dying than in healthy forests.

> *Prediction 13.16*
>
> Nest predation rates by olfactory predators should be lower in forests where many tree species are part of the canopy rather than in forests where a single tree species predominates.

The upper surface of forests composed of multiple species or those with dead trees will be rougher and experience fewer updrafts and turbulence than other forests. Hence, olfactory predators will have a harder time locating nests within them. I am unaware of any studies that have examined these predictions.

> *Prediction 13.17*
>
> Nest predation rates by olfactory predators should be lower in old-growth forests than in younger forests.

Foresters usually define an old-growth forest as one that is overmature or decadent. One key characteristic is the presence of dead trees, which create openings in the canopy where sunlight can penetrate, creating a deep, multilayer crown canopy (Patton 1997). These same characteristics create ideal conditions for airflow above the canopy to penetrate the subcanopy space because of an increase in sweeps and ejections. Hence, ground nests and shrub nests should be safer from olfactory predators in old-growth forests. This prediction remains to be tested.

Impact of edge habitat on olfactory predators

Prediction 13.18

> Nest predation rates (olfactory and visual predators combined) should be lower in forest edge habitat than in forest interior habitat. This decrease should result entirely from lower predation rates by olfactory predators.

Forest edges experience higher levels of turbulence and updrafts than forest interiors. This should make it more difficult for olfactory predators to hunt along forest edges but should not affect visual predators. My prediction that total predation rates should be lower in forest edge habitats is not correct (Paton 1994; Lahti 2001). In fact, many studies have found that predation rates on both artificial nests and nests of free-ranging birds are higher along forest edges than in forest interiors (Wilcove et al. 1986; Andren and Angelstram 1988; Paton 1994; Soderstrom et al. 1998; Batary and Baldi 2004). However, the increased rate of nest predation along forest edges is produced mainly by visual predators rather than by olfactory predators (Angelstam 1986; Paton 1994; Soderstrom et al. 1998).

There seem to be two reasons why nest predation rates are higher along forest edges than in forest interiors. First, there are higher densities of avian nest predators, such as blue jays (*Cyanocitta cristata*), American crows (*Corvus brachyrhynchos*), and common grackles (*Quiscalus quiscula*), along forest edges rather than in forest interiors (Wilcove 1985; Soderstrom et al. 1998; Chalfoun et al. 2002a; Chalfoun et al. 2002b). Second, avian nest predators, such as blue jays, gray jays (*Perisoreus canadensis*), and American crows, prefer to forage along forest edges rather than in forest interiors (Askins 1995; Ibarzabal and Desrochers 2004). Jays also move more slowly and deliberately when hunting along forest edges than in forest interiors (Ibarzabal and Desrochers 2004).

In addition, forest edges attract higher densities of snakes (Chalfoun et al. 2002a) and small mammal predators, such as red squirrels and eastern chipmunks (King et al. 1998), than forest interiors. In a meta-analysis, Chalfoun et al. (2002b) noted that although most studies on predator abundance found no statistically significant difference in predator densities between habitat edges and habitat interiors. Of those studies that did report a difference, however, it was the densities of avian predators along forest edges that differed and not mammalian predators.

All told, it is not surprising that nest predation rates are higher along forest edges than in forest interiors given the high density of nest predators along forest edges. Given this situation, the obvious question is why birds prefer to nest along forest edges.

Prediction 13.19

> The reason why birds prefer to nest along forest edge habitat is that the turbulence and updrafts common along forest edges help hide their nests from olfactory predators.

Early ecologists noted that the highest densities of wildlife occur in edge habitat where two different habitats meet (Leopold 1933). Nesting passerines fit this pattern because a higher density of their nests occurs along forest edges than in the interior of adjacent forests or prairies (Gates and Gysel 1978; Johnson and Temple 1986).

Because nest predation rates are higher along forest edges, why should birds prefer to nest there? Gates and Gysel (1978) believe that birds prefer to nest along forest edges because they contain all of the structural cues birds use to identify safe nesting sites, such as observation perches, cover, and food (hereafter referred to as the *structural cue hypothesis*). Gates and Gysel (1978) and Martin (1992) believe that farming, timber harvesting, and agricultural practices have created unusually narrow forest edges that act as ecological traps. They note that these human-made forest edges have been in existence for only a brief time period relative to the evolutionary time scale, and that birds have not had enough time to adapt to the current situation.

The physics of airflow across forest boundaries provides an alternate explanation for why birds should prefer to nest along forest edges. When compared to forest interiors, forest edges experience more turbulence and higher wind velocities, especially in the subcanopy space. Probably of even greater importance is that forest edges experience updrafts when the wind is moving into the forest from an open field. Hence, it will be harder for olfactory predators to find nests along forest edges than in forest interiors. I hypothesize that birds should respond to this phenomenon by showing a preference for nesting along forest edges to gain protection from olfactory predators (*updraft hypothesis*).

If birds are safer from olfactory predators along forest edges but face an increased risk from visual predators, then why should birds nest there in such high densities? One possibility is that visual predators such as crows and jays eat eggs but do not kill adult birds, while many olfactory predators such as raccoons do. Birds may nest along a forest edge because they are willing to increase their risk of losing their eggs to decrease the risk of losing their own life.

One test of the structural cue and updraft hypotheses about why birds prefer to nest along forest edges is to examine how far into a forest the increased density of nests extends. Edge effects on vegetation generally extend 2*h* into a wood lot (Payne and Bryant 1998); the edge effects on updrafts and turbulence extend 10*h*. Hence, we can test these two hypotheses by assessing whether high nest densities extend farther than the vegetation effect. The results of Gates and Mosher (1981) support the updraft hypothesis. They found that the forest edge was only 13 m wide based on vegetation, but it was 64 m based on an increase in the density of nests. Numerous other investigators have reported that increased nest densities extend over 50 m into the forest from the edge (Paton 1994).

Prediction 13.20

> Ground-nesting or shrub-nesting birds should exhibit greater pref-
> erence for nesting along forest edges than canopy-nesting birds.

Although olfactory predators should experience difficulty detecting and locating prey along a forest edge because of increased turbulence and updrafts, visual predators should not be affected. Hence, ground-nesting or shrub-nesting birds, but not canopy-nesting birds that are already above the reach of most mammalian predators, should prefer to nest along forest edges. The nesting behavior of free-ranging turkey and grouse supports this prediction. These birds nest on the ground and probably have few problems with avian predators because of their size and aggressive behavior. Hence, I predict that olfactory predators should be a much greater threat to nesting turkeys and grouse than visual predators, and thus these birds should show a strong preference for nesting along forest edges. This prediction is supported by the findings that, throughout their range, turkeys select nesting sites along forest edges rather than in forest interiors (Speake et al. 1975; Seiss et al. 1990; Swanson et al. 1996; Thogmartin 1999). Likewise, ruffed grouse prefer to nest less than 10 m from a forest edge or clearing, and those that do so are more successful than grouse that nest away from forest edges (Tirpak et al. 2006).

Evidence against this prediction, however, is provided by Rodewald and Yahner (2001). They examined the nesting success of passerines nesting (that is, wood thrush [*Hylocichla mustelina*] red-eyed vireo [*Vireo olivaceus*], ovenbird [*Seiurus aurocapilla*], and scarlet tanager [*Piranga olivacea*]) in the mature forests of Pennsylvania. They discovered that ground nesters, shrub nesters, and midstory-canopy nesters differed little in how far their nests were located from the forest edge. Nest success also was not related to distance from the forest edge.

Prediction 13.21

> When wood lots or forests are surrounded by open fields, nest pre-
> dation rates by olfactory predators should be lower along the wood
> lot edge facing the wind than on the other three sides.

More updrafts and turbulence will occur along the windward side of a wood lot than along two sides that are parallel to the wind or on the leeward side of the wood lot. I am unaware of any data that either support or refute this prediction.

chapter fourteen

Using the physics of airflow to redefine common ecological terms

Ecologists have generally viewed and defined the world based on light. This approach is valid in that plant distribution is related to light intensity, and herbivore distribution is related to plant distribution. Yet, not everything is related to light and vision. Olfactory predators and their prey live in a parallel world of olfaction where everything is related to airflow and turbulence. In this chapter, I redefine common ecological terms utilizing this alternate view of the world. In doing so, my goal is not to convince ecologists to redefine their terms but rather to promote the reality that human reliance on vision influences our perception of the world, our definitions of the natural world, and how we classify objects. If olfaction were the dominant sense of humans, our perceptions, definitions, and classification schemes would be different.

My second goal is to demonstrate a weakness in some of our basic definitions of ecology; many basic terms are ambiguous and lack operational definitions. Consequently, two scientists can look at the same landscape and disagree on what they are viewing, or two ecologists can map the different habitats and their edges within a 1-km² piece of ground and produce maps that vary considerably. To elucidate the problems caused by having ambiguous terms, I use some examples from forest ecology, and then I illustrate how these same terms can be operationally defined based on how an olfactory predator would perceive them.

Examples from forest ecology of the confusion that can be created by ambiguous definitions

Decline of populations of neotropical migrants owing to forest fragmentation have caused concern. Hence, it is worthwhile examining the conflicting definitions that are used by scientists exploring if and why neotropical migrants are declining.

What is a forest patch or habitat patch?

A *habitat patch* is commonly defined as an area that is different from its surroundings (Kotliar and Wiens 1990; Thogmartin 1999). However, this definition provides no guidance about what different means or how large a piece of habitat must be before it qualifies as a habitat patch. Not surprisingly, scientists differ in their criteria for what would qualify as a unique habitat patch.

What is a forest interior?

Many authors have argued that the reason large contiguous tracts of forest need to be maintained is that only they contain unique forest interior habitat that is essential for some avian species. *Forest interior habitat* is usually defined as habitat a certain distance away from a forest edge, although authors vary considerably regarding where forest interior habitat begins relative to the forest edge. Forest interior habitat was believed to start 25 m from an edge by Sargent et al. (1998), 60 m from an edge by Thogmartin (1999), 250 m by Donovan et al. (1997), and 400 m by Heske (1995). Laurance (1991) argues that edge effects extend 500 m into tropical forests.

What is a forest edge?

One hypothesis about why predation rates are higher in small wood lots than in large ones involves the ratio of edge habitat to interior habitat; this ratio increases as forest stands become smaller and predation rates along forest edges are particularly high. Before this hypothesis can be tested, we need first to define a forest edge. Again, lack of consensus exists on what constitutes a forest edge. Many researchers limit their study of edge to outward forest edges where forests end and large tracts of open habitat begin (Angelstam 1986); other researchers consider a forest edge to include any change in habitat type (Thogmartin (1999). For the latter group, the boundary between different forest cover types would constitute a forest edge (for example, wherever a pine and deciduous forest adjoin). Other authors, who tested whether predation rates were higher along forest edges, limited their research to "artificial edges" created by human activity such as timber harvest, agriculture, or development (Paton 1994). For example, Donovan et al. (1997) assume that forest edges occur where forests adjoin agricultural fields, abandoned agricultural fields in early successional plant communities, or forest openings wider than 300 m and made by humans.

How far does a forest edge extend into a forest?

Many scientists define the width of a forest edge based on where enhanced nest predation rates occur. The rationale is that a forest edge extends into the forest until a point is reached where predation rates are similar to those occurring within the interior of the forest. Wilcove et al. (1986) suggest that the edge-related increase in predation levels off at 200 to 500 m from the edge; Andren and Angelstam (1988), Paton (1994), and Batary and Baldi (2004) showed that predation rates were high only within 50 m of the edge. Sargent et al. (1998) consider edge habitat to be less than 10 m from a forest edge.

What is a forest clearing?

My search of the scientific literature has not revealed an operational definition of a forest clearing; moreover, there is debate over how big a hole in the forest canopy must be before it qualifies as a clearing. Perhaps the best answer is the simplest: It must be big enough to create edge habitat. Brittingham and Temple (1983) consider 0.25-ha openings in the forest canopy to create edge habitat, but Donovan et al. (1997) assume that forest clearings have to be more than 300 m wide. Thogmartin (1999) considers any road running through a forest as a clearing, regardless of the road's width; predictably, many other researchers disagree. Paton (1994) recognizes this problem and proposes that a clearing would create edge habitat if its width were three times the height of the forest.

Benefits of defining ecological terms based on the physics of airflow

Ecologists usually try to determine questions about habitat and landscape by looking for changes in the plant community. From the standpoint of an olfactory predator, however, the plants that make up a habitat are unimportant. What is important is the habitat's roughness and how it affects the movement of air. Thus, all habitats can be defined in terms of their zero-plane displacement (d) and aerodynamic roughness length (z_0). These terms are dependent on the h and λ of surface features within the habitat patch along with the density and distribution of the surface features. In this approach, two habitats can be operationally defined as different from each other if their values of h and λ are different. This can be determined through sampling and the use of statistics. Ecotones and their widths can be defined because the profile of the wind changes whenever it crosses two surfaces of different roughness. The width of the habitat edge can be defined as the distance over which the fetch extends back from the habitat's edge in both directions.

One advantage of using the physics of airflow to define ecological terms is that the definitions are then scalable across different habitats. The same definitions that are applied to forests are applied to grasslands and other open habitats. The commonly accepted definitions of many ecological terms are not easily scalable.

Examples are given next of some common ecological terms that have been defined purely on the basis of the physics of airflow. The terms are not placed in alphabetical order but rather are grouped by concept. For instance, all terms dealing with habitat patches or types of habitat are grouped together as are all terms dealing with habitat edges.

Habitat is a contiguous area with the same surface roughness (i.e., with similar values of d and z_0).

Habitat patch is habitat that is large enough to allow air flowing over it to adjust to the new surface roughness of the habitat patch. Generally, the width of a habitat patch has to be greater than $20h$. The operational definition is that it is an area large enough that the profile of airflow over it has enough space to stabilize before exiting the habitat patch.

Habitat boundary is the place where one habitat patch ends and another begins.

Edge habitat is that part of a habitat patch close enough to a habitat boundary that air flowing over it has not had time to adjust to the surface roughness of the habitat patch. Therefore, it is a region along the edge of a habitat patch that experiences higher levels of turbulence and more updrafts than the habitat interior. Edge habitat usually extends from the habitat boundary into the habitat patch for a distance of $10h$. The operational definition is that it is an area along the boundary of a habitat patch where (1) the surface boundary layer has significantly increased in height over that found in the habitat interior or (2) where turbulence (σ_u, σ_v, or σ_w) is significantly higher than in the habitat interior.

Interior habitat is that part of the habitat patch far enough removed from the habitat boundary that air flowing over it has had time to adjust to the surface roughness of the habitat patch. Interior habitat occurs at distances beyond $10h$ of a habitat boundary. The operational definition that it is the part of a habitat patch that is far enough away from the habitat edge so that mean height of the surface boundary layer and turbulence (σ_u, σ_v or σ_w) have become constant.

Middle-interior habitat is the outer section of interior habitat where subcanopy airflow is reduced owing to a lack of airflow coming from the direction of a habitat edge. The reason for this is because a breeze entering the forest from a forest edge is deflected upward by updrafts occurring within the edge habitat. This is along a zone 10 to $50h$ from a habitat's edge. Middle-interior habitat is operationally defined as interior habitat where the mean direction of airflow significantly deviates from the direction of airflow occurring in the boundary layer above it.

Deep-interior habitat is the inner section of interior habitat where airflow within the subcanopy space is no longer influenced by the habitat edge. This is at distances greater than $50h$ of a habitat's edge. Deep-interior habitat is operationally defined as interior habitat where the mean direction of airflow within the subcanopy space is similar to the direction of airflow occurring in the boundary layer above it.

Habitat clearing is a small area within a habitat patch that differs in surface roughness from the rest of the habitat patch. Generally, habitat clearings must be greater than $1h$ when their surface features are higher than the surrounding habitat patch and greater than $3h$ when their surface features are less than the surrounding habitat patch. The operational definition of a habitat clearing is that it is a contiguous area sharing similar values of d and z_0 that is (1) completely surrounded by a habitat patch having a different d and z_0 and (2) large enough in size to cause an increase in turbulence (σ_u, σ_v, or σ_w) but too small to be a habitat patch.

Old-growth habitat is a habitat where d and z_0 no longer change over time.

Dominant surface features are those that extend above the plant canopy and act as isolated surface features.

Dominant plants are plants that are tall enough to be dominant surface features.

Forest habitat is a habitat patch where the dominant plants are dense enough to create a subcanopy space and high enough that its subcanopy space is taller than a person of average height (1.8 m). As a rule of thumb, a habitat patch will qualify as a forest patch when most plants are higher than 2.6 m and dense enough that the canopy of the taller plants covers at least 25% of the surface area. The operational definition of forest habitat is that it is a habitat patch where the vegetation is thick enough and high enough that d is greater than 1.8 m.

Brush habitat is operationally defined as a habitat patch where dominant plants are dense enough to create a subcanopy space and where d ranges in height from 0.5 to 1.8 m. Generally, a habitat patch will qualify as a brush patch when the height of the plant canopy ranges from 0.7 to 2.6 m, and this canopy covers at least 25% of the surface area.

Open habitat is a habitat patch where the vegetation or other surface features are short enough to create a subcanopy space d ranging in height from 0.0 to 0.5 m. The operational definition is the same as the common definition. As a rule of thumb, a habitat patch will qualify as an open patch when the plant canopy is less than 0.7 m high.

Grass-dominated habitat is an open habitat patch where the canopy is composed of grass or other flexible plants that bend and orient in the direction of the wind (when wind velocities are of moderate intensity). Because grass leaves are angled upward, air is deflected upward when it strikes them. Grass-dominated habitat's operational definition is an open habitat where 50% or more of the dominant plants orient their canopies in a stream-wise direction when wind velocities exceed 20 km/hour. This causes airflow in the subcanopy space to be angled upward (>0°).

Forb-dominated habitat is an open habitat patch where the canopy is composed of forbs or other plants with stiff leaves and stems that will not bend and orient in the direction of wind even when wind velocities are of moderate intensity. Mean airflow in the subcanopy space of forb-dominated habitat is horizontal. The operational definition is that it is an open habitat where 50% or more of the dominant plants do not orient their canopies in a stream-wise direction when wind velocities exceed 20 km/hour and therefore do not reduce their drag by more than 30%.

Prairie is the same as an open habitat.

Savanna habitat is an open habitat patch where surface features are far enough apart that they cause considerable turbulence close to the ground, resulting in an increase in z_0 but not in d. Savanna habitat occurs when isolated surface features occupy 0.1 to 5% of the surface area.

chapter fifteen

Epilogue

I was motivated to write this book because I realize that humans are visual predators, and that our perceptions of the world are based on this one modality. We are cognizant of how animals hide from predators that use vision to hunt them. Yet, I wanted to know how animals hide from predators that use olfaction to hunt them. My curiosity about a simple question grew into a theory that animals can and do hide from olfactory predators, and that there is a constant struggle between olfactory predators and the animals that hide from them. I called this the *olfactory concealment theory*. I realized not only that animals face risks from both their depositional odor trails and odor plumes, but also that animals can take measures to reduce the risk that each poses to them.

Dangers posed by depositional odor trails

Every time an animal walks, its risk of falling victim to an olfactory predator increases; that is, by moving, the animal lays down a depositional odor trail that informs olfactory predators not only if the animal's passage, but also how long ago it was there and in which direction it was heading. Olfactory predators can then follow the depositional odor trail to stalk the animal and kill it. An animal can reduce the strength of its depositional odor trail by grooming itself constantly, especially its feet. An animal can make it harder for olfactory predators to follow its depositional odor trail by joining with other conspecifics so that its trail intersects others, by traveling long distances quickly so that a part of its depositional odor trail will have time to disappear before the predator reaches that part of the trail, by crossing water bodies, or by moving only when it is sunny. The behavior of animals trailed by olfactory predators has rarely been reported in the scientific literature, although deer, rabbits, and other animals often exhibit these escape behaviors when chased by dogs.

Dangers posed by odor plumes

All animals, even stationary ones, constantly create an odor plume that poses a much greater danger to them than depositional odor trails because animals cannot stop emitting airborne odorants. The ability of olfactory predators to detect and locate prey using odor plumes varies with time, location, and atmospheric conditions. These variances result from updrafts that cause odor plumes to rise higher than a predator's olfactory zone and atmospheric turbulence that makes odor plumes meander. Addditionally, both lateral and vertical turbulence cause odor plumes to disperse rapidly and shortens the distance over which an olfactory predator can use them to detect prey. The optimal wind velocity for olfactory predators is about 5 to 20 km/h; suboptimal wind speeds make it difficult for predators to detect and locate their quarry using odor plumes as do winds that are too fast.

Thanks to the work of meteorologists, we know that updrafts and both vertical and lateral turbulence vary with the time of day, atmospheric stability, differential heating of the earth's surface, wind velocities, and frictional forces against features on the earth's surface. The time of day and locations of updrafts and turbulence are based on the physics of airflow and are largely predicted in both time and space. Meteorologists' work has shown that convective currents exist where there is differential heating of the earth's surface by the sun. Such heating occurs when adjacent surfaces differ in their ability to absorb or conduct heat (e.g., when water bodies are adjacent to dry land) and when one surface receives more sunlight than another because they differ in aspect and slope or because one surface is more shaded than another (e.g., the ground surface beneath a forest vs. a prairie or bare surface). Research has demonstrated that animals that are vulnerable to predators take advantage of the updrafts created by warmer surfaces to hide from olfactory predators.

Updrafts and turbulence are also caused by frictional forces when airflow is blocked by features on the ground such as hills and boulders, isolated trees and bushes, habitat edges, and shelterbelts. Updrafts and turbulence are produced by rough plant canopies and ground surfaces that have high aerodynamic roughness length z_0. The olfactory concealment theory states that animals should hide from olfactory predators by positioning themselves in these areas, and those individuals that do so will have a higher probability of surviving than those that hide elsewhere. In many but not all cases, research has shown that fawns, calves, and nesting birds are found in higher densities on or near surface features and rough surfaces, and that nest predation rates by olfactory predators are lower when nests are located in areas where turbulence and updrafts are prevalent.

The data in this book demonstrate clearly that animals live in a complex world where there are not only olfactory predators but also visual and auditory predators. In some cases, optimal hiding places from olfactory and visual predators are the same, but often this is not the case. Hence, animals make trade-offs between the competing needs to avoid detection by visual and olfactory predators. They also make numerous trade-offs between their need to hide from olfactory predators and their other needs to survive and reproduce. To maximize their inclusive fitness, animals must determine the optimal solution to the complex problem of where to hide or nest but cannot do so because they have only imperfect knowledge. Furthermore, the optimal solution to this problem is constantly changing owing to decisions that other animals make about where to hide, atmospheric conditions, alternate prey densities and locations, relative densities of different types of predators, and the current search images and hunting tactics of those predators. For these reasons, studies that have tried to identify which nest site characteristics are correlated with nest densities and nesting success have produced conflicting results. After all, predators are also trying to identify these same characteristics so that they can learn where to hunt. If they succeed, those optimal nest site characteristics will not be optimal for long. Hence, one truth that I hope I demonstrated in this book is that animals have great difficulty trying to survive in a world when some predators are seeking them using olfaction while others are using vision or sound.

Can the olfactory-concealment theory help guide future research and provide answers to questions that heretofore have lacked explanation?

Any good theory should be able to guide future research; by that standard, the olfactory concealment theory is a useful one because Chapters 11 through 14 are full of predictions that have yet to be tested. A good theory should be able to answer questions that heretofore have lacked explanation. I think the olfactory concealment theory passes this test. I reiterate a few examples here.

1. Why should saliva contain digestive enzymes? The digestive enzymes in saliva have almost no time to digest food before they are destroyed by the stomach so their function has been much debated. The olfactory concealment theory offers an explanation: Digestive enzymes in saliva are useful when grooming to destroy odorants on the animal's surface.

2. Why not have multiple entrances to a cavity nest? Two entrances to a nest cavity could allow an incubating bird to escape when a predator appears at one of them. Yet, nest cavities only have one entrance. The olfactory concealment theory provides an explanation: Cavities with only one entrance produce an intermittent odor plume that olfactory predators have difficulty following.

3. Why are nest densities high along forest edges when nest predation rates are also high there? The olfactory concealment theory offers an explanation: Forest edges create updrafts, so birds nesting there are safe from olfactory predators but not from visual predators. However, many olfactory predators are midsize mammals that kill incubating adults; most visual predators are ravens, crows, and jays that just take eggs. For most birds, losing a clutch of eggs is better than being killed.

4. Why do old-growth forests offer better nesting habitat for some birds than even-aged forests? The olfactory concealment theory points out that the upper surface of an old-growth forest is much rougher than an even-aged forest, and because of this, there are more sweeps and ejections into the subcanopy of old-growth forests.

5. Why do birds prefer to nest and why do monkeys like to sleep on branches that hang over water? The olfactory concealment theory provides an explanation: Water bodies are warmer than ground surfaces at night and create updrafts. Hence, the odor plumes of animals nesting or sleeping above them will be carried above the olfactory zone of predators on the ground.

6. Why do pregnant ungulates seek south-facing slopes to give birth, and why do fawns prefer to hide there? The olfactory concealment theory points out that updrafts are prevalent on south-facing slopes. Because of these updrafts, predators that are at a lower elevation than the ungulate will not be able to detect it using olfaction.

Does the olfactory-concealment theory have any applied value?

Ultimately, a sound theory must have applied value. In this case, the olfactory concealment theory can help guide the efforts of wildlife managers and conservationists. Three examples follow.

1. Ducks Unlimited, the U.S. Fish and Wildlife Service, and many private landowners manage land for upland-nesting ducks by creating a uniform stand of dense nesting cover. The olfactory concealment theory argues that to protect nests from olfactory predators, however, the goal should be to increase the habitat's z_0. This means that rather than creating a solid and smooth canopy of tall plants, the objective should be to create variance in plant height so that an area has numerous pockets of tall shrubs, short shrubs, tall grasses, short grasses, and various forbs.

2. Even-aged forests provide poor protection for birds nesting on the ground or in the subcanopy because few sweeps are directed downward owing to the smoothness of the forest canopy. The olfactory concealment theory points out that to

improve nest success, the upper surface of the forest canopy should be made rougher by thinning or encouraging canopy trees of different heights. Old-growth or multiage forests experience more sweeps into the subcanopy space and along the ground; therefore, better protection is provided from olfactory predators.

3. The olfactory concealment theory shows that when holes are drilled in trees to create cavity nests or when nest boxes are erected for birds such as the endangered red-cockaded woodpecker, these should be built on either the windward or leeward side of the tree. This will help hide the nest from olfactory predators because airflow out of such a nest will be more intermittent than from nests that are located on the two sides of a tree that are perpendicular to the wind's direction.

References

Albone, E.S., P.E. Gosden, and G.C. Ware, 1977, Bacteria as a source of chemical signals in mammals, in D. Muller-Schwarze and M.M. Mozell, Eds., *Chemical Signals in Vertebrates*, Plenum Press, New York, pp. 35–43.

Albrecht, T., D. Horak, J. Kreisinger, K. Weidinger, P. Klvana, and T.C. Michot, 2006, Factors determining pochard nest predation along a wetland gradient, *Journal of Wildlife Management* 70:784–791.

Alcock, J, 2001, *Animal Behavior*, 7th ed., Sinauer Associates, Sunderland, MA.

Aldeman, S., C.R. Taylor, and N.C. Heglund, 1975, Sweating on paws and palms: what is its function? *American Journal of Physiology* 229:1400–1402.

Aldridge, C.L. and R.M. Brigham, 2002, Sage-grouse nesting and brood habitat use in southern Canada, *Journal of Wildlife Management* 66:433–444.

Alldredge, A.W., R.D. Beblinger, and J. Peterson, 1991, Birth and fawn bed site selection by pronghorns in the sagebrush-steppe community, *Journal of Wildlife Management* 55:222–227.

Amiro, B.D., 1990, Comparison of turbulence statistics within three boreal forest canopies, *Boundary-Layer Meteorology* 51:99–121.

Anderson, J.R., 1984, Ethology and ecology of sleep in monkeys and apes, *Advances in the Study of Behavior* 14:165–229.

Andren, H., 1992, Corvid density and nest predation in relation to forest fragmentation: a landscape perspective, *Ecology* 73:794–804.

Andren, H. and P. Angelstam, 1988, Elevated predation rates as an edge effect in habitat islands: experimental evidence, *Ecology* 69:544–547.

Angelstam, P., 1986, Predation on ground-nesting birds' nests in relation to predator densities and habitat edge, *Oikos* 47:365–373.

Arnold, T.W., M.D. Sorenson, and J.J. Rotella, 1993, Relative success of overwater and upland mallard nests in southwestern Manitoba, *Journal of Wildlife Management* 57:578–581.

Askins, R.A., 1995, Hostile landscapes and the decline of migratory songbirds, *Science* 267:1956–1957.

Atema, J., 1995, Chemical signals in the marine environment: dispersal, detection, and temporal signal analysis, *Proceedings of the National Academy of Science* 92:62–66.

Austin, G.T., 1974, Nesting success of the cactus wren in relation to nest orientation, *Condor* 76:216–217.

Autenrieth, R.E. and E. Fichter, 1975, On the behavior and socialization of pronghorn fawns, *Wildlife Monographs* 42:1–111.

Axel, R., 1995, The molecular logic of smell, *Scientific American* 273:154–159.

Aylor, D.E., Y. Wang, and D.R. Miller, 1993, Intermittent wind close to the ground within a grass canopy, *Boundary-Layer Meteorology* 66:427–448.

Badyaev, A.V., 1995, Nesting habitat and nesting success of eastern wild turkeys in the Arkansas Ozark highlands, *Condor* 97:221–232.

Badyaev, A.V., T.E. Martin, and W.J. Etges, 1996, Habitat sampling and habitat selection of female wild turkeys: ecological correlates and reproductive consequences, *Auk* 113:636–646.

Baines, D., 1996, The implications of grazing and predator management on the habitats and breeding success of black grouse *Tetrao tetrix*, *Journal of Applied Ecology* 33:54–62.

Baldocchi, D.D. and T.P. Meyers, 1988, Turbulence structure in a deciduous forest, *Boundary-Layer Meteorology* 43:345–364.

Baldocchi, D.D., S.B. Verma, and N.J. Roserberg, 1983, Characteristics of airflow above and within soybean canopies, *Boundary-Layer Meteorology* 25:43–54.

Balgooyen, T.C., 1990, Orientation of American kestrel nest cavities: revisited, *Journal of Raptor Research* 24:27–28.

Ballard, W.B., H.A. Whitlaw, S.J. Young, R.A. Jenkins, and G.J. Forbes, 1999, Predation and survival of white-tailed deer fawns in northcentral New Brunswick, *Journal of Wildlife Management* 63:574–579.

Barka, T., 1980, Biologically active polypeptides in submandibular glands, *Journal of Histochemistry and Cytochemistry* 28:836–859.

Barratt, R., 2001, *Atmospheric Dispersion Modelling: An Introduction to Practical Applications*, Earthscan Publications, Sterling, VA.

Barrett, M.W., 1981, Environmental characteristics and functional significance of pronghorn fawn bedding sites in Alberta, *Journal of Wildlife Management* 45:120–131.

Barrette, C. and F. Messier, 1980, Scent-marking in free-ranging coyotes (*Canis latrans*), *Animal Behaviour* 28:814–819.

Batary, P. and A. Baldi, 2004, Evidence of an edge effect on avian nest success, *Conservation Biology* 18:389–400.

Batchelor, R.F., 1965, *The Roosevelt Elk in Alaska: Its Ecology and Management*, Alaska Department of Fish and Game, Juneau, AK.

Bednekoff, P.A. and A.I. Houston, 1994, Avian daily foraging patterns: effects of digestive constraints and variability, *Evolutionary Ecology* 8:36–52.

Bekesy, G., 1964, Olfactory analogue to directional hearing, *Journal of Applied Physiology* 19:369–373.

Bekoff, M., A.C. Scott, and D.A. Conner, 1989, Ecological analyses of nesting success in evening grosbeaks, *Oecologia* 81:67–74.

Bergen, J.D., 1975, Air movement in a forest clearing as indicated by smoke drift, *Agricultural Meteorology* 15:165–179.

Berger, J., 1991, Pregnancy incentives, predation constraints and habitat shifts: experimental and field evidence for wild bighorn sheep, *Animal Behaviour* 41:61–77.

Bergerud, A.T., 1985, Antipredator strategies of caribou: dispersion along shorelines, *Canadian Journal of Zoology* 63:1324–1329.

Bergerud, A.T., H.E. Butler, and D.R. Miller, 1984, Antipredator tactics of calving caribou: dispersion in mountains, *Canadian Journal of Zoology* 62:1566–1575.

Bergerud, A.T. and M.W. Gratson, 1988, Survival and breeding strategies of grouse, in A.T. Bergerud and M.W. Gratson, Eds., *Adaptive Strategies and Population Ecology of Northern Grouse*, University of Minnesota Press, Minneapolis, MN, pp. 473–577.

Best, L.B., 1978, Field sparrow reproductive success and nesting ecology, *Auk* 95:9–22.

Best, L.B. and D.F. Stauffer, 1980, Factors affecting nesting success in riparian bird communities, *Condor* 82:149–158.

Blackadar, A.K., H.A. Panofsky, P.E. Glass, and J.F. Boogaard, 1967, Determination of the effect of roughness change on the wind profile, *The Physics of Fluids* (Supplement) 209–211.

Bleich, V.C., 1999, Mountain sheep and coyotes: patterns of predator evasion in a mountain ungulate, *Journal of Mammalogy* 80:283–289.

Bleich, V.C., R.T. Bowyer, and J.D. Wehausen 1997, Sexual segregation in mountain sheep: resources or predation? *Wildlife Monographs* 134:1–50.

Bollinger, E.K. and T.A. Gavin, 2004, Responses of nesting bobolinks (*Dolichonyx oryzivorus*) to habitat edges, *Auk* 121:767–776.

Borgo, J.S., L.M. Conner, and M.R. Conover, 2006, Role of predator odor in roost site selection of southern flying squirrels, *Wildlife Society Bulletin* 34:144–149.

Bossert, W.H. and E.O. Wilson, 1963, The analysis of olfactory communication among animals, *Journal of Theoretical Biology* 5:443–469.

Bowman, G.B. and L.D. Harris, 1980, Effect of spatial heterogeneity on ground nest depredation, *Journal of Wildlife Management* 44:806–813.

Bowyer, R.T., J.G. Kie, and V. Van Ballenberghe, 1998, Habitat selection by neonatal black-tailed deer: climate, forage, or risk of predation? *Journal of Mammalogy* 79:415–425.

Bradley, E.F., 1968, A micrometeorological study of velocity profiles and surface drag in the region modified by a change in surface roughness, *Quarterly Journal of the Royal Meteorological Society* 94:361–379.

Bradley, E.F., 1980, An experimental study of the profiles of wind speed, shearing stress, and turbulence at the crest of a large hill, *Quarterly Journal of the Royal Meteorological Society* 106:101–123.

Bradley, R.M., 1991, Salivary secretion, in T.V. Getchell, R.L. Doty, L.M. Bartoshuk, and J.B. Snow, Jr., Eds., *Smell and Taste in Health and Disease*, Raven Press, New York, pp. 127–144.

Brady, J., G. Gibson, and M.J. Packer, 1989, Odour movement, wind direction, and the problem of host-finding by tsetse flies, *Physiological Entomology* 14:369–380.

Brady, J., N. Griffiths, and Q. Paynter, 1995, Wind speed effects on odor source location by tsetse flies (*Glossina*), *Physiological Entomology* 20:293–302.

Brighton, P.W.M., 1978, Strongly stratified flow past three-dimensional obstacles, *Quarterly Journal of the Royal Meteorological Society* 104:289–307.

Brittingham, M.C. and S.A. Temple, 1983, Have cowbirds caused forest songbirds to decline? *BioScience* 33:31–35.

Brua, R.B., 1999, Ruddy duck nesting success: do nest characteristics deter nest predation? *Condor* 101:867–870.

Budgett, H.M., 1933, *Hunting by Scent*, Charles Scribner's Sons, New York.

Bump, G., R.W. Darrau, F.C. Edminster, and W.F. Crissey, 1947, *The Ruffed Grouse: Life History, Propagation, Management*, Holling, Buffalo, NY.

Burghardt, G.M., 1966, Stimulus control of the prey attack response in naïve garter snakes, *Psychonomic Science* 4:37–38.

Burghardt, G.M. and E.H. Hess, 1968, Factors influencing the chemical release of prey attack in newborn snakes, *Journal of Comparative and Physiological Psychology* 66:289–295.

Burtt, E.H., Jr. and J.M. Ichida, 2004, Gloger's rule, feather-degrading bacteria, and color variation among song sparrows, *Condor* 106:681–686.

Byers, S.M., 1974, Predator–prey relationships on an Iowa waterfowl nesting area, *Transactions of the North American Wildlife and Natural Resources Conference* 39:223–229.

Cain, J.W., III, K.S. Smallwood, M.L. Morrison, and H.L. Loffland, 2006, Influence of mammal activity on nesting success of passerines, *Journal of Wildlife Management* 70:522–531.

Calladine, J., D. Baines, and P. Warren, 2002, Effects of reduced grazing on population density and breeding success of black grouse in northern England, *Journal of Applied Ecology* 39:772–780.

Cannon, S.K. and F.C. Bryant, 1997, Bed-site characteristics of pronghorn fawns, *Journal of Wildlife Management* 61:1134–1141.

Carbone, C., J.T. Du Toit, and I.J. Gordon, 1997, Feeding success in African wild dogs: does kleptoparasitism by spotted hyenas influence hunting group size? *Journal of Animal Ecology* 66:318–326.

Carbone, C., L. Frame, G. Frame, J. Malcolm, J. Fanshawe, C. FitzGibbon, G. Shaller, I.J. Gordon, J.M. Rowcliffe, and J.T. Du Toit, 2005. Feeding success of African wild dogs (*Lycaon pictus*) in the Serengeti: the effects of group size and kleptoparasitism, *Journal of Zoology, London* 266:153–161.

Carde, R.T., 1986, Epilogue: behavior mechanisms, in T.L. Payne, M.C. Birch, and C.E.J. Kennedy, Eds., *Mechanisms in Insect Olfaction*, Clarendon Press, Oxford, U.K., pp. 174–186.

Caro, T., 2005, *Antipredator Defenses in Birds and Mammals*, University of Chicago Press, Chicago.

Carroll, B.K. and D.L. Brown, 1977, Factors affecting neonatal fawn survival in southern-central Texas, *Journal of Wildlife Management* 41:63–69.

Caughey, S.J., J.C. Wyngaard, and J.C. Kaimal, 1979, Turbulence in the evolving stable boundary layer, *Journal of Atmospheric Sciences* 36:1041–1052.

Chalfoun, A.D., M.J. Ratnaswamy, F.R. Thompson III, 2002a, Songbird nest predators in forest-pasture edge and forest interior in a fragmented landscape, *Ecological Applications* 12:858–867.

Chalfoun, A.D., F.R. Thompson III, and M.J. Ratnaswamy, 2002b, Nest predators and fragmentation: a review and meta-analysis, *Conservation Biology* 306–318.

Chesness, R.A., M.M. Nelson, and W.H. Longley, 1968, The effect of predator removal on pheasant reproductive success, *Journal of Wildlife Management* 32:683–697.

Christensen, T.A., T. Heinbockel, and J.G. Hildebrand, 1996, Olfactory information processing in the brain: encoding chemical and temporal features of odors, *Journal of Neurobiology* 30:82–91.

Clapperton, B.K., E.O. Minot, and D.R. Crump, 1988, An olfactory recognition system in the ferret *Mustela fero* L. (Carnivora: Mustelidae), *Animal Behaviour* 36:541–553.

Clapperton, B.K., E.O. Minot, and D.R. Crump, 1989, Scent lures from the anal sac secretions of the ferret (*Mustela furo* L.), *Journal of Chemical Ecology* 15:291–308.

Clark, C.W. and D.A. Levy, 1988, Diel vertical migrations by pelagic plankivorous fishes and the antipredator window, *American Naturalist* 131:271–290.

Clark, L. and P.S. Shah, 1992, Information content of prey odor plumes: what do foraging Leach's storm-petrels know? in R.L. Doty and D. Muller-Schwarze, Eds., *Chemical Signals in Vertebrates*, Plenum Press, New York, pp. 421–428.

Clark, R.G. and T.D. Nudds, 1991, Habitat patch size and duck nesting success: the crucial experiments have not been performed, *Wildlife Society Bulletin* 19:534–543.

Clark, R.G. and B.K. Wobeser, 1997, Making sense of scents: effects of odor on survival of simulated duck nests, *Journal of Avian Biology* 28:31–37.

Clifford, R.J., 1964, Some notes and theories on "scent." Its formation, properties, and usage as derived from observations on and experience with tracker and patrol dogs in Malaya, *Veterinary Annual* 10–21.

Connelly, J.W., M.A. Schroeder, A.R. Sands, and C.E. Braun, 2000, Guidelines to manage sage grouse populations and their habitats, *Wildlife Society Bulletin* 28:967–985.

Conner, R.N., 1975, Orientation of entrances to woodpecker nest cavities, *Auk* 92:371–374.

Conover, M.R., 1987, Acquisition of predator information by active and passive mobbers in ring-billed gull colonies, *Behaviour* 102:41–57.

Conover, M.R., 2001, *Resolving Human–Wildlife Conflicts: The Science of Wildlife Damage Management*, Lewis Brothers, Boca Raton, FL.

Conover, M.R. and K.S. Lyons, 2005, Will free-ranging predators stop depredating untreated eggs in pulegone-scented gull nests after exposure to pulegone-injected eggs, *Applied Animal Behaviour* 93:135–145.

Conover, M.R. and D.E. Miller, 1978, Reaction of ring-billed gulls to predators and human disturbances at their breeding colonies, *Colonial Waterbirds* 2:41–47.

Cook, R.S., M. White, D.O. Trainer, and W.C. Glazener, 1971, Mortality of young white-tailed deer fawns in south Texas, *Journal of Wildlife Management* 35:47–56.

Cowardin, L.M., D.S. Gilmer, and C.W. Shaiffer, 1985, Mallard recruitment in the agricultural environment of North Dakota, *Wildlife Monographs* 92:1–37.

Crabtree, R.L., L.S. Broome, and M.L. Wolfe, 1989, Effects of habitat characteristics on gadwall nest predation and nest-site selection, *Journal of Wildlife Management* 53:129–137.

Creel, S. and N.M. Creel, 2002, *The African Wild Dog*, Princeton University Press, Princeton, NJ.

Crockett, A.B. and H.H. Hadow, 1975. Nest site selection by Williamson and red-naped sapsuckers, *Condor* 77:365–368.

Czaplicki, J.A. and R.H. Porter, 1974, Visual cues mediating the selection of goldfish (*Carassius auratus*) by two species of *Natrix*, *Journal of Herpetology* 8:129–134.

David, C.T., 1986, Mechanisms of directional flight in wind, in T.L. Payne, M.C. Birch, and C.E.S. Kennedy, Eds., *Mechanisms in Insect Olfaction*, Clarendon Press, Oxford, U.K., pp. 49–57.

David, C.T., J.S. Kennedy, and A.R. Ludlow, 1983, Finding of a sex pheromone source by gypsy moths released in the field, *Nature* 303:804–806.

David, C.T., J.S. Kennedy, A.R. Ludlow, J.N. Perry, and C. Wall, 1982, A reappraisal of insect flight toward a distant point source of wind-borne odour, *Journal of Chemical Ecology* 8:1207–1215.

Davidson, W.B. and E. Bollinger, 2000, Predation rates on real and artificial nests of grassland birds, *Auk* 117:147–153.

DeGraaf, R.M., 1995, Nest predation rates in managed and reserved extensive northern hardwood forests, *Forest Ecology and Management* 79:227–234.

Dennis, J.V., 1971, Species using red-cockaded woodpecker holes in northeastern South Carolina, *Bird-banding* 42:79–87.

Deshmukh, I., 1986, *Ecology and Tropical Biology,* Blackwell Scientific Publications, Palo Alto, CA.

DeVos, M., F. Parte, J. Railed, P. Laffort, and L.J. Van Geert, 1990, *Standardized Human Olfactory Thresholds,* Oxford University Press, Oxford, U.K.

DiBenedetto, C., 2005, Beagle, *Field and Stream* 110(6):48.

Dimmick, R.L. and A.B. Akers, 1969, *An Introduction to Experimental Aerobiology,* John Wiley and Sons, New York.

Donovan, T.M., P.W. Jones, E.M. Annand, and F.R. Thompson III, 1997, Variation in local-scale edge effects: mechanisms and landscape context, *Ecology* 78:2064–2075.

Doving, K.B., 1990, Scent trailing by tracking dogs. What is the physiological basis for concentration coding? in D. Child, Ed., *Chemosensory Information Processing,* Springer-Verlag, New York, pp. 271–276.

Drummond, H.M., 1979, Stimulus control of amphibious predation in the northern water snake *Nerodia s. sipedon, Zeitschrift fur Tierpsychologie* 50:18–44.

Duebbert, H.F., 1969, High nest density and hatching success of ducks on South Dakota CAP land, *Transactions of the North American Wildlife and Natural Resource Conference* 34:218–229.

Duebbert, H.F., J.T. Lokemoen, and D.E. Sharp, 1983, Concentrated nesting of mallards and gadwalls on Miller Lake Island, North Dakota, *Journal of Wildlife Management* 47:729–740.

Elkinton, J.S., C. Schal, T. Ono, and R.T. Carde, 1987, Pheromone puff trajectory and upwind flight of male gypsy moths in a forest, *Physiological Entomology* 12:399–406.

Ellison, S.A., 1979, The identification of salivary components, in I. Kleinberg, S.A. Ellison, and I.D. Mandel, Eds., *Saliva and Dental Caries,* Special Supplement to Microbiology Abstracts, Informational Retrieval, New York, pp. 13–29.

Erikstad, K.E., R. Blom, and S. Myrberget, 1982, Territorial hooded crows as predators on willow ptarmigan nests, *Journal of Wildlife Management* 46:109–114.

Erlinge, S., 1977, Stacing strategy in stoat *Mustela erminea, Oikos* 28:32–42.

Espmark, Y. and R. Langvatn, 1979, Cardiac responses in alarmed red deer calves, *Behavioral Processes* 4:179–188.

Espmark, Y. and R. Langvatn, 1985, Development and habituation of cardiac and behavioral responses in young red deer calves (*cervus elaphus*) exposed to alarm stimuli, *Journal of Mammology* 66:702–711.

Estrada, A. and R. Estrada, 1976, Establishment of a free ranging colony of stump-tailed macaques (*Macaca arctoides*): relations to the ecology, I, *Primates* 17:337–355.

Fanshawe, J.H. and C.D. Fitzgibbon, 1993, Factors influencing the hunting success of an African wild dog pack, *Animal Behaviour* 45:479–490.

Farbman, A.I., 2000, Cell biology of olfactory epithelium, in T.E. Finger, W.L. Silver, and D. Restrepo, Eds., *The Neurobiology of Taste and Smell,* 2nd ed., Wiley-Liss, New York, pp. 131–158.

Festa-Bianchet, M., 1988, Seasonal range selection in bighorn sheep: conflicts between forage quality, forage quantity, and predator avoidance, *Oecologia* 75:580–586.

Filliater, T.S., R. Breitwisch, and P.M. Nealen, 1994, Predation on northern cardinal nests: does choice of nest site matter? *Condor* 96:761–768.

Fitzjarrald, D.E., 1973, A field investigation of dust devils, *Journal of Applied Meteorology* 12:808–813.

Fox, K.B. and P.R. Krausman, 1994, Fawning habitat of desert mule deer, *Southwestern Naturalist* 39:269–275.

Gabrielsen, G.W., A.S. Blix, and H. Ursin, 1985, Orienting and freezing responses in incubating ptarmigan hens, *Physiology and Behavior* 34:925–934.

Garratt, J.R., 1977, *Aerodynamic Roughness and Mean Monthly Surface Stress over Australia,* CSIRO Division of Atmospheric Physics Technical Paper 29, Canberra, Australia.

Garratt, J.R., 1980, Surface influence upon vertical profiles in the atmospheric near-surface layer, *Quarterly Journal of the Royal Meteorological Society* 106:803–819.

Garratt, J.R., 1992, *The Atmospheric Boundary Layer,* Cambridge University Press, Cambridge, U.K.

Gash, J.H.C., 1986, Observations of turbulence downwind of a forest-heath interface, *Boundary-Layer Meteorology* 36:227–237.

Gates, J.E. and L.W. Gysel, 1978, Avian nest dispersion and fledging success in field-forest ecotones, *Ecology* 59:871–883.

Gates, J.E. and J.A. Mosher, 1981, A functional approach to estimating habitat edge width for birds, *American Midland Naturalist* 105:189–192.

Gautier-Hion, A., 1973, Social and ecological features of talapoin monkey: comparisons with sympatric Cercopithecines, in R.P. Michael and J.H. Crook, Eds., *Comparative Ecology and Behaviour of Primates*, Academic Press, New York, pp. 147–170.

Gawlik, D.E., M.E. Hostetler, and K.L. Bildstein 1988, Naphthalene moth balls do not deter mammalian predators at red-winged blackbird nests, *Journal of Field Ornithology* 59:189–191.

Geist, V., 2002, Adaptive behavioral strategies, in D.E. Toweill and J.W. Thomas, Eds., *North American Elk: Ecology and Management*, Smithsonian Institute Press, Washington, DC, pp. 389–433.

Gendron, R.P., 1986, Searching for cryptic prey: evidence for optimal search rates and the formation of search images in quail, *Animal Behaviour* 34:898–912.

Gendron, R.P. and J.E.R. Staddon, 1983, Searching for cryptic prey: the effect of search rate, *American Naturalist* 121:172–185.

Gibbs, J.P., 1991, Avian nest predation in tropical wet forest: an experimental study, *Oikos* 60:155–161.

Gilbert, B.K., 1972, The influence of foster rearing on adult social behavior in fallow deer (*Dama dama*), in V. Geist and F. Walther, Eds., *Symposium on the Behaviour of Ungulates and Its Relation to Management*, IUCN Publications 24, Morges, Switzerland, pp. 247–273.

Giroux, J.F., 1981, Use of artificial islands by nesting waterfowl in southeastern Alberta, *Journal of Wildlife Management* 45:669–679.

Gorman, M.L., 1984, The response of prey to stoat (*Mustela erminea*) scent, *Journal of Zoology, London* 202:419–423.

Gotmark, F., D. Blomqvist, O.C. Johansson, and J. Bergkvist, 1995, Nest site selection: a trade-off between concealment and view of the surroundings, *Journal of Avian Biology* 26:305–312.

Grant, T.A., E.M. Madden, T.L. Shaffer, P.J. Pietz, G.B. Berkey, and N.J. Kadrmas, 2006, Nest survival of clay-colored and vesper sparrows in relation to woodland edge in mixed-grass prairies, *Journal of Wildlife Management* 70:691–701.

Graul, W.D., 1975, Breeding biology of the mountain plover, *Wilson Bulletin* 87:6–31.

Green, G.A. and R.G. Anthony, 1989, Nesting success and habitat relationships of burrowing owls in the Columbia Basin, Oregon, *Condor* 91:347–354.

Green, S.R., J. Grace, and N.J. Hutchings, 1995, Observations of turbulent airflow in three stands of widely spaced Sitka spruce, *Agricultural and Forest Meteorology* 74:205–225.

Greenwood, P.J., 1980, Mating systems, philopatry, and dispersal in birds and mammals, *Animal Behaviour* 28:1140–1162.

Greenwood, R.J., A.B. Sargeant, D.H. Johnson, L.M. Cowardin, and T.L. Shaffer, 1995, Factors associated with duck nest success in the Prairie Pothole region of Canada, *Wildlife Monographs* 128:1–57.

Gregory, P.H., 1973, *The Microbiology of the Atmosphere*, 2nd ed., John Wiley and Sons, New York.

Griffiths, N. and J. Brady, 1995, Wind structure in relation to odour plumes in tsetse fly habitats, *Physiological Entomology* 20:286–292.

Grubb, T.C., 1972, Smell and foraging in shearwaters and petrels, *Nature* 237:404–405.

Grubb, T.C., 1973, Colony location by Leach's petrel, *Auk* 90:78–82.

Grubb, T.C., 1974, Olfactory navigation to the nesting burrow in Leach's petrel (*Oceanodroma leucorrhoa*), *Animal Behaviour* 22:192–202.

Guilford, T. and M.S. Dawkins, 1987, Search images not proven: a reappraisal of recent evidence, *Animal Behaviour* 35:1838–1845.

Gustafsson, L. and S.G. Nilsson, 1984, Clutch size and breeding success of pied and collared flycatchers *Ficedula* spp. in nest-boxes of different sizes, *Ibis* 127:380–385.

Guthery, F.S., W.P. Meinzer, Jr., S.L. Beasom, and M. Caroline, 1984, Evaluation of placed baits for reducing coyote damage in Texas, *Journal of Wildlife Management* 48:621–626.

Guthery, F.S., A.R. Rybak, S.D. Fuhlendorf, T.L. Hiller, S.G. Smith, W.H. Puckett, Jr., and R.A. Baker, 2005, Aspects of the thermal ecology of bobwhites in North Texas, *Wildlife Monographs* 159:1–36.

Gutzwiller, K.J., 1990, Minimizing dog-induced biases in game bird research, *Wildlife Society Bulletin* 18:351–356.

Hammond, M.C. and W.R. Forward, 1956, Experiments on causes of duck nest predation, *Journal of Wildlife Management* 20:243–247.

Hansell, M.H., 2000, *Bird Nests and Construction Behaviour*, Cambridge University Press, Cambridge, U.K.

Hardy, P.C. and M.L. Morrison, 2001, Nest site selection by elf owls in the Sonoran desert, *Wilson Bulletin* 113:23–32.

Hargrove, J.W. and G.A. Vale, 1978, The effect of host odor concentration on catches of tsetse flies Glossinidae and other Diptera in the field, *Bulletin of Entomological Research* 68:607–612.

Harmeson, J.P., 1974, Breeding ecology of the diskcissel, *Auk* 91:348–359.

Harper, J.A., J.H. Harn, W.W. Bentley, and C.F. Yocom, 1967, The status and ecology of the Roosevelt elk in California, *Wildlife Monographs* 16:1–49.

Harvey, P.H., P.J. Greenwood, and C.M. Perrins, 1979, Breeding area fidelity of great tits (*Parus major*), *Journal of Animal Ecology* 48:305–313.

Hass, C.A., 1998, Effects of prior nesting success on site fidelity and breeding dispersal: an experimental approach, *Auk* 115:929–936.

Heisler, G.M. and D.R. Dewalle, 1988, 2. Effects of windbreak structure on wind flow, *Agriculture, Ecosystems and Environment* 22/23:41–69.

Hepper, P.G. and D.L. Wells, 2005, How many footsteps do dogs need to determine the direction of an odour trail? *Chemical Senses* 30:291–298.

Herlugson, C.J., 1981, Nest site selection in mountain bluebirds, *Condor* 83:252–255.

Hernandez, F., S.E. Henke, N.J. Silvy, and D. Rollins, 2003, The use of prickly pear cactus as nesting cover by northern bobwhites, *Journal of Wildlife Management* 67:417–423.

Heske, E.J., 1995, Mammalian abundances on forest-farm edges versus forest interiors in southern Illinois: is there an edge effect? *Journal of Mammalogy* 76:562–568.

Hicks, B.B., P. Hyson, and C.J. Moore, 1975, A study of eddy fluxes over a forest, *Journal of Applied Meteorology* 14:58–66.

Hill, D.A., 1984, Factors affecting nest success in the mallard and tufted duck, *Ornis Scandinavica* 15:115–122.

Hirth, D.H., 2000, Behavioral ecology, in S. Demarais and P.R. Krausman, Eds., *Ecology and Management of Large Mammals in North America*, Prentice Hall, Upper Saddle River, NJ, pp. 175–191.

Holloran, M.J., A.G. Lyon, S.J. Slater, J.L. Kuipers, and S.H. Anderson, 2005, Greater sage-grouse nesting habitat selection and success in Wyoming, *Journal of Wildlife Management* 69:638–649.

Honza, M., I.J. Oien, A. Moksnes, and E. Roskaft, 1998, Survival of reed warbler *Acrocephalus scirpaceus* clutches in relation to nest position, *Bird Study* 45:104–108.

Houston, D.B., 1982, *The Northern Yellowstone Elk Herd*, Macmillan, New York.

Howard Hughes Medical Institute, 2005, *Serving Science: '05 Annual Report*, Chevy Chase, MD.

Howard, M.N., S.K. Skagen, and P.L. Kennedy, 2001, Does habitat fragmentation influence nest predation in the shortgrass prairie? *Condor* 103:530–536.

Huegul, C.N., 1985, Predator-Avoidance Behaviors in White-Tailed Deer that Favor Fawn Survival, Ph.D. dissertation, Iowa State University, Ames.

Huegel, C.N., R.B. Dahlgren, and H.L. Gladfelter, 1986, Bed site selection by white-tailed deer fawns in Iowa, *Journal of Wildlife Management* 50:474–480.

Hunt, J.C.R., 1980, Wind over hills, in J.C. Wyngaard, Ed., *Workshop of the Planetary Boundary Layer*, American Meteorological Society, Boston, pp. 107–244.

Hurst, G.A., L.W. Burger, and B.D. Leopold, 1996, Predation and Galiforme recruitment: an old issue revisited, *Transactions of the North American Wildlife and Natural Resources Conference* 61:62–76.

Ibarzabal, J. and A. Desrochers, 2004, A nest predator's view of managed forest: gray jay (*Perisoreus canadensis*) movement patterns in response to forest edges, *Auk* 121:162–169.

Inouye, D.W., 1976, Nonrandom orientation of entrance holes to woodpecker nests in aspen trees, *Condor* 78:101–102.

Inouye, R.S., N.J. Huntly, and D.W. Inouye, 1981, Non-random orientation of gila woodpecker nest entrances in saguaro cacti, *Condor* 83:88–89.

Irvine, M.R., B.A. Gardiner, and M.K. Hill, 1997, The evolution of turbulence across a forest edge, *Boundary-Layer Meteorology* 84:467–496.

Jackson, R.M., M. White, and F.F. Knowlton, 1972, Activity patterns of young white-tailed deer fawns in South Texas, *Ecology* 53:262–270.

Jackson, S.L., D.S. Hik, and R.F. Rockwell, 1988, The influence of nesting habitat on reproductive success of the lesser snow goose, *Canadian Journal of Zoology* 66:1699–1703.

Jacobsen, N.K., 1979, Alarm bradycardia in white-tailed deer fawns (*Odocoileus virginianus*), *Journal of Mammalogy* 60:343–349.

Janzen, D.H., 1978, Predation intensity on eggs on the ground in two Costa Rican forests, *American Midland Naturalist* 100:467–470.

Jarvis, P.G., G.B. James, and J.J. Landsberg, 1976, Coniferous forest, in J.L. Monteith, Ed., *Vegetation and the Atmosphere*, Vol. 2, Academic Press, New York, pp. 171–240.

Jimenez, J.E., 1999, Nest Success of Dabbling Ducks in a Human-Modified Prairie: Effects of Predation and Habitat Variables at Different Spatial Scales, Ph.D. dissertation, Utah State University, Logan.

Jimenez, J.E. and M.R. Conover 2001, Ecological approaches to reduce predation on ground-nesting gamebirds and their nests, *Wildlife Society Bulletin* 29:62–69.

Johnson, D.E., 1951, Biology of the elk calf, *Cervus canadensis nelsoni*, *Journal of Wildlife Management* 15:396–410.

Johnson, G.R., 1977, *Tracking Dog: Theory and Method*, 2nd ed., Arner Publications, Westmoreland, NY.

Johnson, R.G. and S.A. Temple, 1986, Assessing habitat quality for birds nesting in fragmented tallgras prairies, in J. Verner, M.L. Morrison, and C.J. Ralph, Eds., *Wildlife 2000: Modeling Habitat Relationships of Terrestrial Vertebrates*, University of Wisconsin Press, Madison, pp. 245–249.

Johnson, R.G. and S.A. Temple, 1990, Nest predation and brood parasitism of tallgrass prairie birds, *Journal of Wildlife Management* 54:106–111.

Jones, C.D., 1983, On the structure of instantaneous plumes in the atmosphere, *Journal of Hazardous Materials* 7:87–112.

Kaimal, J.C. and J.A. Businger, 1970, Case studies of a convective plume and a dust devil, *Journal of Applied Meteorology* 9:612–620.

Kaissling, K.E., 1997, Pheromone-controlled anemotaxis in moths, in M. Lehrer, Ed., *Orientation and Communication in Arthropods*, Birkhauser Verlag, Basel, Switzerland, pp. 343–374.

Kalmus, H., 1955, The discrimination by the nose of the dog of individual human odors and in particular of the odors of twins, *British Journal of Animal Behaviour* 3:25–31.

Karg, G., D.M. Sucking, and S.J. Bradley, 1994, Absorption and release of pheromone of *Epiphyas postvittana* (Lepidoptera: Tortricidae) by apple leaves, *Journal of Chemical Ecology* 20:1825–1841.

Keith, L.B., 1961, A study of waterfowl ecology on small impoundments in southeastern Alberta, *Wildlife Monographs* 6:1–88.

Kennedy, J.S., 1983, Zigzagging and casting as a programmed response to wind-borne odour: a review, *Physiological Entomology* 8:109–120.

Kennedy, J.S., A.R. Ludlow, and C.J. Sanders, 1981, Guidance of flying male moths by wind-borne sex pheromone, *Physiological Entomology* 6:395–412.

Keough, H.L., 2006, Factors Influencing Breeding Ferruginous Hawks (*Buteo regalis*) in the Uintah Basin, Utah, PhD. dissertation, Utah State University, Logan.

King, D.I., C.R. Griffin, and R.M. DeGraaf, 1998, Nest predator distribution among clearcut forest, forest edge and forest interior in an extensively forested landscape, *Forest Ecology and Management* 104:151–156.

Klopfer, P.H. and J.U. Ganzhorn, 1985, Habitat selection: behavioral aspects, in M. Cody, Ed., *Habitat Selection in Birds*, Academic Press, New York, pp. 435–453.

Komar, D., 1999, The use of cadaver dogs in locating scattered, scavenged human remains: preliminary field test results, *Journal of Forensic Science* 44:405–408.

Kondo, J. and H. Yamazawa, 1986, Aerodynamic roughness over an inhomogeneous ground surface, *Boundary-Layer Meteorology* 35:331–348.

Kotliar, N.B. and J.A. Wiens, 1990, Multiple scales of patchiness and patch structure: a hierarchical framework for the study of heterogeneity, *Oikos* 59:253–260.

Krasowski, T.P. and T.D. Nudds, 1986, Microhabitat structure of nest sites and nesting success of diving ducks, *Journal of Wildlife Management* 50:203–208.

Krausman, P.R. and D.M. Shackleton, 1999, Bighorn sheep, in S. Demarais and P.R. Krausman, *Ecology and Management of Large Mammals in North America*, Prentice Hall, Upper Saddle River, NJ, pp. 517–544.

Krestel, D., D. Passe, J.C. Smith, and L. Jonsson, 1984, Behavioral determination of olfactory thresholds to amyl acetate in dogs, *Neuroscience and Biobehavioral Reviews* 8:169–174.

Kruuk, H., 1972, *The Spotted Hyena*, University of Chicago Press, Chicago.

Kunkel, K.E. and L.D. Mech, 1994, Wolf and bear predation on white-tailed deer fawns in northeastern Minnesota, *Canadian Journal of Zoology* 72:1557–1565.

Lahti, D.C., 2001, The "edge effect on nest predation" hypothesis after 20 years, *Biological Conservation* 99:365–374.

Laing, D.G., 1991, Characteristics of the human sense of smell when processing odor mixtures, in D.G. Laing, R.L. Doty, and W. Breipohl, Eds., *The Human Sense of Smell*, Springer-Verlag, New York, pp. 241–259.

Langen, T.A., D.T. Bolger, and T.J. Case, 1991, Predation on artificial bird nests in chaparral fragments, *Oecologia* 86:395–401.

Lanyon, W.E., 1957, *The Comparative Biology of the Meadowlarks* (Sturnella*) in Wisconsin*, Nuttall Ornithological Club, Cambridge, MA.

Lariviere, S. and F. Messier, 1998, Effect of density and nearest neighbours on simulated waterfowl nests: can predators recognize high-density nesting patches? *Oikos* 83:12–20.

Laurance, W.F., 1991, Edge effects in tropical forest fragments: application of a model for the design of nature reserves, *Biological Conservation* 57:205–219.

Leader, N. and Y. Yom-Tov, 1998, The possible function of stone ramparts at the nest entrance of the blackstart, *Animal Behaviour* 56:207–217.

Leclerc, M.Y., K.C. Beissner, R.H. Shaw, G.D. Hartog, and H.H. Neumann, 1991, The influence of buoyancy on third-order turbulence velocity statistics with a deciduous forest, *Boundary-Layer Meteorology* 55:109–123.

Lee, X., 2000, Air motion within and above forest vegetation in non-ideal vegetation in non-ideal conditions, *Forest Ecology and Management* 135:3–18.

Lee, X. and T.A. Black, 1993, Turbulence near the forest floor of an old-growth Douglas-fir stand on a south-facing slope, *Forest Science* 39:211–230.

Lent, P.C., 1974, Mother-infant relationships in ungulates, in V. Geist and F. Walther, Eds., *Symposium on the Behaviour of Ungulates and Its Relation to Management*, IUCN Publications 24, Morges, Switzerland, pp. 14–55.

Leonard, M.L. and J. Picman, 1987, Nesting mortality and habitat selection by marsh wrens, *Auk* 104:491–495.

Leopold, A., 1933, *Game Management*, Charles Scribner and Sons, New York.

Lequette, B., C. Verheyden, and P. Jouventin, 1989, Olfaction in subantarctic sea birds: its phylogenetic and ecological significance, *Condor* 91:732–735.

Lettau, H., 1969, Note on aerodynamic roughness-parameter estimation on the basis of roughness-element description, *Journal of Applied Meteorology* 8:828–832.

Lima, S.L., 1988a, Initiation and termination of daily feeding in dark-eyed juncos: influences of predation risk and energy reserves, *Oikos* 53:3–11.

Lima, S.L., 1988b, Vigilance during the initiation of daily feeding in dark-eyed juncos, *Oikos* 53:12–16.

Lindgren, P.M.F., T.P. Sullivan, and D.R. Crump, 1997, Review of synthetic predator odor semiochemicals as repellents for wildlife management in the Pacific Northwest, in J.R. Mason, Ed., *Repellents in Wildlife Management*, U.S. Department of Agriculture, Wildlife Services, National Wildlife Research Center, Fort Collins, CO, pp. 217–230.

Lingle, S., 2002, Coyote predation and habitat segregation of white-tailed deer and mule deer, *Ecology* 83:2037–2048.

Linnell, J.D.C., E.B. Nilsen, and R. Andersen, 2004, Selection of bed-sites by roe deer (*Capreolus capreolus*) fawns in an agricultural landscape, *Acta Theriologica* 49:103–111.

Linsdale, J.M. and P.Q. Tomich, 1953, *A Herd of Mule Deer*, University of California Press, Berkeley.

Livezey, B.C., 1981, Duck nesting in retired cropland at Horicon National Wildlife Refuge, Wisconsin, *Journal of Wildlife Management* 45:27–37.

Lokemoen, J.T. and T.A. Messmer, 1994, *Locating, Constructing and Managing Islands for Nesting Waterfowl*, Jack Berryman Institute Publication 5, Utah State University, Logan.

Lokemoen, J.T. and R.O. Woodward, 1993, An assessment of predator barriers and predator control to enhance duck nest success on peninsulas, *Wildlife Society Bulletin* 21:275–282.

Lowe, B., 1981, *Hunting the Clean Boot*, Blandford Press, Poole, U.K.

Lusk, J.J., S.G. Smith, S.D. Fuhlendorf, and F.S. Guthery, 2006, Factors influencing northern bobwhite nest-site selection and fate, *Journal of Wildlife Management* 70:564–571.

Lyons, T.J. and W.D. Scott, 1990, *Principles of Air Pollution Meteorology*, CRC Press, Boca Raton, FL.

Mankin, P.C. and R.E. Warner, 1992, Vulnerability of ground nests to predation on an agricultural habitat island in east-central Illinois, *American Midland Naturalist* 128:281–291.

Manzer, D.L. and S.J. Hannon, 2005, Relating grouse nest success and corvid density to habitat: a multi-scale approach, *Journal of Wildlife Management* 69:110–123.

Mao, J.S., M.S. Boyce, D.W. Smith, F.J. Singer, D.J. Vales, J.M. Vore, and E.H. Merrill, 2005, Habitat selection by elk before and after wolf reintroduction in Yellowstone National Park, *Journal of Wildlife Management* 69:1691–1707.

Marini, M.A. and C. Melo, 1998, Predators of quail eggs, and the evidence of the remains: implications for nest predation studies, *Condor* 100:395–399.

Marples, M.J., 1969, Life on the human skin, *Scientific American* 220:108–115.

Marshall, D.A. and D.G. Moulton, 1981, Olfactory sensitivity to alpha-ionone in humans and dogs, *Chemical Senses* 6:53–61.

Martin, T.E., 1987, Artificial nest experiments: effects of nest appearance and type of predator, *Condor* 89:925–928.

Martin, T.E., 1992, Breeding productivity considerations: what are the appropriate habitat features for management? in J.M. Hagan III and D.W. Johnston, Eds. *Ecology and Conservation of Neotropical Migrant Landbirds*, Smithsonian Institution Press, Washington, DC, pp. 455–473.

Martin, T.E., 1993, Nest predation and nest sites: new perspectives on old patterns, *BioScience* 43:523–532.

Marzluff, J.M. 1988, Do pinyon jays alter nest placement based on prior experience? *Animal Behaviour* 36:1–10.

Mason, J.R. and L. Clark, 2000, The chemical senses in birds, in G.A. Whittow, Ed., *Sturkie's Avian Physiology*, 5th ed., Academic Press, New York, pp. 39–56.

Maxson, S.J. and M.R. Riggs, 1996, Habitat use and nest success of overwater nesting ducks in westcentral Minnesota, *Journal of Wildlife Management* 60:108–119.

McBean, G.A., 1968, An investigation of turbulence within the forest, *Journal of Applied Meteorology* 7:410–416.

McCartney, W., 1968, *Olfaction and Odours: An Osphresiological Essay*, Springer-Verlag, New York.

McKee, G., M.R. Ryan, and L.M. Mechlin, 1998, Predicting greater prairie-chicken nest success from vegetation and landscape characteristics, *Journal of Wildlife Management* 62:314–321.

McKinnon, D.T. and D.C. Duncan, 1999, Effectiveness of dense nesting cover for increasing duck production in Saskatchewan, *Journal of Wildlife Management* 63:382–389.

McNaughton, K.G., 1988, 1. Effects of windbreaks on turbulent transport and microclimate, *Agriculture, Ecosystems and Environment* 22/23:17–39.

McNider, R.T. and R.A. Pielke, 1984, Numerical simulation of slope and mountain flows, *Journal of Climate and Applied Meteorology* 23:1441–1453.

Mech, L.D., 1966, *The Wolves of Isle Royale*, Fauna of the National Parks of the United States, Fauna Series 7, Department of the Interior, National Park Service, Washington, DC.

Mech, L.D., 1970, *The Wolf: The Ecology and Behavior of an Endangered Species*, American Museum of Natural History, Natural History Press, Garden City, NY.

Mech, L.D., 1984, Predators and predation, in L.K. Halls, Ed., *White-Tailed Deer: Ecology and Management*, Stackpole Books, Harrisburg, PA, pp. 189–200.

Mertens, J.A.L., 1977, Thermal conditions for successful breeding in great tits (*Parus major* L.), *Oecologia* 28:31–56.

Miller, D.R., J.D. Lin, and Z.N. Lu, 1991, Air flow across an alpine forest clearing: a model and field measurements, *Agricultural and Forest Meteorology* 56:209–225.

Miller, H.W., 1971, Relationships of duck nesting success to land use in North and South Dakota, *Transactions of the Congress of the International Union of Game Biologists* 10:133–141.

Miller, L., 1942, Some tagging experiments with black-footed albatrosses, *Condor* 44:3–9.

Mills, M.G.L., 1990, *Kalahari Hyaenas*, Unwin Hyman, London.

Moen, A.N., M.A. DellaFera, A.L. Hiller, and B.A. Buxton, 1978, Heart rates of white-tailed deer fawns in response to recorded wolf howls, *Canadian Journal of Zoology* 56:1207–1210.

Moller, A.P., 1988, Nest predation and nest site choice in passerine birds in habitat patches of different size: a study of magpies and blackbirds, *Oikos* 53:215–221.

Moran, D.T., B.W. Jafek, and J.C. Rowley III, 1991, The ultrastructure of the human olfactory muscosa, in D.G. Laing, R.L. Doty, and W. Breipohl, Eds., *The Human Sense of Smell*, Springer-Verlag, New York, pp. 1–28.

Mozell, M.M., P.F. Kent, P.W. Scherer, D.E. Hornung, and S.J. Murphy, 1991, Nasal airflow, in T.V. Getchell, R.L. Doty, L.M. Bartoshuk, and J.B. Snow, Jr., Eds., *Smell and Taste in Health and Disease*, Raven Press, New York, pp. 481–492.

Muller-Schwarze, D., 1972, Response of young black-tailed deer to predator odors, *Journal of Mammalogy* 53:393–394.

Munn, R.E., 1966, *Descriptive Micrometeorology*, Academic Press, New York.

Murie, A., 1944, *The Wolves of Mount McKinley*, Fauna of the National Parks of the United States, Fauna Series Number 5, Washington, DC.

Murlis, J., 1997, Odor plumes and the signal they provide, in R.T. Carde and A.K. Minks, Eds., *Insect Pheromone Research: New Directions*, Chapman and Hall, New York, pp. 221–231.

Murlis, J., J.S. Elkinton, and R.T. Carde, 1992, Odor plumes and how insects use them, *Annual Review of Entomology* 37:505–532.

Nakamura, K., 1976, The effect of wind velocity on the diffusion of *Spodoptera litura* (F.) sex pheromone, *Applied Entomology and Zoology* 11:312–319.

Nams, V.O., 1991, Olfactory search images in striped skunks, *Behaviour* 119:267–284.

Nams, V.O., 1997, Density-dependent predation by skunks using olfactory search images, *Oecologia* 110:440–448.

Nef, P., 1993, Early events in olfaction: diversity and spatial patterns of odorant receptors, *Receptors and Channels* 1:259–266.

Neff, W.D. and C.W. King, 1987, Observations of complex terrain flows using acoustic sounders: drainage flow structure and evolution, *Boundary-Layer Meteorology* 43:15–41.

Nelson, M.E. and L.D. Mech, 1986, Mortality of white-tailed deer in northeastern Minnesota, *Journal of Wildlife Management* 50:691–698.

Nelson, T.A. and A. Woolf, 1987, Mortality of white-tailed deer fawns in southern Illinois, *Journal of Wildlife Management* 51:326–329.

Nevitt, G., 1999, Foraging by sea birds on an olfactory landscape, *American Scientist* 87:46–53.

Nilsson, S.G., 1984, The evolution of nest-site selection among hole-nesting birds: the importance of nest predation and competition, *Ornis Scandinavica* 15:167–175.

Nilsson, S.G., 1986, Evolution of hole-nesting in birds: on balancing selection pressures, *Auk* 103:432–435.

Nilsson, S.G., K. Johnsson, and M. Tjernberg, 1991, Is avoidance by black woodpeckers of old nest holes due to predators? *Animal Behaviour* 41:439–441.

Noble, G.K. and H.J. Clausen, 1936, The aggregation behavior of *Storeria dekayi* and other snakes with especial reference to the sense organs involved, *Ecological Monographs* 6:269–316.

Nolan, V., Jr., 1978, *The Ecology and Behavior of the Prairie Warbler* Dendroica discolor, Ornithological Monographs 26, American Ornithological Union, Lawrence, KS.

Nord, M., 1991, Shelter effects of vegetation belts—results of field measurements, *Boundary-Layer Meteorology* 54:363–385.

Norment, C.J., 1993, Nest-site characteristics and nest predation in Harris' sparrows and white-crowned sparrows in the Northwest Territories, Canada, *Auk* 110:769–777.

Oniki, Y., 1979, Is nesting success of birds low in the tropics? *Biotropica* 11:60–69.

Ozoga, J.J., L.J. Verme, and C.S. Bienz, 1982, Parturition behavior and territoriality in white-tailed deer: impact on neonatal mortality, *Journal of Wildlife Management* 46:1–11.

Palmer, W.E., S.R. Priest, R.S. Seiss, P.S. Phalen, and G.A. Hurst, 1993, Reproductive effort and success in a declining wild turkey population, *Proceedings of the Annual Conference of the Southeastern Association of Fish and Wildlife Agencies* 47:138–147.

Pasitschniak-Arts, M. and F. Messier, 1995, Risk of predation on waterfowl nests in the Canadian prairies: effects of habitat edges and agricultural practices, *Oikos* 73:347–355.

Pasquill, F., 1961, The estimation of the dispersion of windborne material, *Meteorological Magazine* 90:33–49.

Passe, D.H. and J.C. Walker, 1985, Odor psychophysics in vertebrates, *Neuroscience Biobehavioral Review* 9:431–467.

Paton, P.W.C., 1994, The effect of edge on avian nesting success: how strong is the evidence? *Conservation Biology* 8:17–26.

Patton, D.R., 1997, *Wildlife Habitat Relationships in Forested Ecosystems*, rev. ed., Timber Press, Portland, OR.

Payne, N.F. and F.C. Bryant, 1998, *Wildlife Habitat Management on Forestlands, Rangelands, and Farm-lands*, Krieger Publishing, Malabar, FL.

Perry, J.N. and C. Wall, 1986, The effect of habitat on the flight of moths orienting to pheromone sources, in T.L. Payne, M.C. Birch, and C.E.J. Kennedy, Eds., *Mechanisms in Insect Olfaction*, Clarendon Press, Oxford, U.K., pp. 91–96.

Petit, K.E., L.J. Petit, and D.R. Petit, 1989, Fecal sac removal: do the pattern and distance of dispersal affect the chance of nest predation? *Condor* 91:479–482.

Phillips, T.A., 1974, Characteristics of elk calving habitat on the Sawtooth National Forest, *Range Improvement Notes* 19:1–5.

Picman, J., 1988, Experimental study of predation on eggs of ground-nesting birds: effects of habitat and nest distribution, *Condor* 90:124–131.

Pierre, J.P., H. Bears, and C.A. Paskowski, 2001, Effects of forest harvesting on nest predation in cavity-nesting waterfowl, *Auk* 118:224–230.

Pietrewicz, A.T. and A.C. Kamil, 1979, Search image formation in the blue jay (*Cyanocitta cristata*), *Science* 204:1332–1333.

Plate, E.J., 1971, The aerodynamics of shelter belts, *Agricultural Meteorology* 8:203–222.

Preston, F.W., 1957, The look-out perch as a factor in predation by crows, *Wilson Bulletin* 69:368–370.

Quader, S., 2006, What makes a good nest? Benefits of nest choice to female Baya weavers (*Ploceus philippinus*), *Auk* 123:475–486.

Rachlow, J.L. and R.T. Bowyer, 1998, Habitat selection by Dall's sheep (*Ovis dalli*): maternal trade-offs, *Journal of Zoology (London)* 245:457–465.

Ramakrishnan, U. and R.G. Coss, 2001, Strategies used by bonnet macaques (*Macaca radiata*) to reduce predation risk while sleeping, *Primates* 42:193–206.

Rands, M.R.W., 1982, *The Influence of Habitat on the Population Ecology of Partridges*, Ph.D. dissertation, Oxford University, Oxford, U.K.

Rands, M.R.W., 1986, Effect of hedgerow characteristics on partridge breeding densities, *Journal of Applied Ecology* 23:479–487.

Rangen, S.A., R.G. Clark, and K.A. Hobson, 2000, Visual and olfactory attributes of artificial nests, *Auk* 117:136–146.

Raphael, M.G., 1985, Orientation of American kestrel nest cavities and nest trees, *Condor* 87:437–438.

Raupach, M.R., 1994, Simplified expressions for vegetation roughness length and zero-plane displacement as functions of canopy height and area index, *Boundary-Layer Meteorology* 71:211–216.

Raupach, M.R., J.J. Finnigan, and Y. Brunet, 1996, Coherent eddies and turbulence in vegetation canopies: the mixing-layer analogy, *Boundary-Layer Meteorology* 78:351–382.

Rawson, N.E., 2000, Human olfaction, in T.E. Finger, W.L. Silver, and D. Restrepo, Eds., *The Neuro-biology of Taste and Smell*, 2nd ed., Wiley-Liss, New York, pp. 257–284.

Raynor, G.S., 1971, Wind and temperature structure in a coniferous forest and a contiguous field, *Forest Science* 17:351–363.

Redmond, G.W., D.M. Keppie, and P.W. Herzog, 1982, Vegetative structure, concealment, and success at nests of two races of spruce grouse, *Canadian Journal of Zoology* 60:670–675.

Regnier, F.E. and M. Goodwin, 1977, On the chemical and environmental modulation of pheromone release from vertebrate scent marks, in D. Muller-Schwarze and M.M. Mozell, Eds., *Chemical Signals in Vertebrates*, Plenum Press, New York, pp. 115–133.

Reifsnyder, W.E., G.M. Furnival, and J.L. Horowitz, 1971, Spatial and temporal distribution of solar radiation beneath forest canopies, *Agricultural Meteorology* 9:21–37.

Riley, T.Z., C.A. Davis, M. Ortiz, and M.J. Wisdom, 1992, Vegetative characteristics of successful and unsuccessful nests of lesser prairie chickens, *Journal of Wildlife Management* 56:383–387.

Roberts, S.D., J.M. Coffey, and W.F. Porter, 1995, Survival and reproduction of female wild turkeys in New York, *Journal of Wildlife Management* 59:437–447.

Roberts, S.D. and W.F. Porter, 1998a, Influence of temperature and precipitation on survival of wild turkey poults, *Journal of Wildlife Management* 62:1499–1505.

Roberts, S.D. and W.F. Porter, 1998b, Relation between weather and survival of wild turkey nests, *Journal of Wildlife Management* 62:1492–1498.

Rodewald, A.D. and R.H. Yahner, 2001, Avian nesting success in forested landscapes: influence of landscape composition, stand and nest-patch microhabitat, and biotic intereactions, *Auk* 118:1018–1028.

Rosenberry, J.L. and W.D. Klimstra, 1970, The nesting ecology and reproductive performance of the eastern meadowlark, *Wilson Bulletin* 82:243–267.

Sargeant, A.B. and L.E. Eberhardt, 1975, Death feigning by ducks in response to predation by red foxes (*Vulpes fulva*), *American Midland Naturalist* 94:108–119.

Sargent, R.A., J.C. Kilgo, B.R. Chapman, and K.V. Miller, 1998, Effect of stand width and adjacent habitat on breeding bird communities in bottomland hardwoods, *Journal of Wildlife Management* 62:72–83.

Schal, C., 1982, Intraspecific vertical stratification as a mate-finding mechanism in tropical cockroaches, *Science* 215:1405–1407.

Schranck, B.W., 1972, Waterfowl nest cover and some predation relationships, *Journal of Wildlife Management* 36:182–186.

Schwede, G., H. Hendrichs, and C. Wemmer, 1992, Activity and movement patterns of young white-tailed deer fawns, in R.D. Brown, Ed., *The Biology of Deer*, Springer-Verlag, New York, pp. 56–62.

Segal, M., Y. Mahrer, and R.A. Pielke, 1982, Application of a numerical mesoscale model for the evaluation of seasonal persistent regional climatological patterns, *Journal of Applied Meteorology* 21:1754–1762.

Seiss, R.S., P.S. Phalen, and G.A. Hurst, 1990, Wild turkey nesting habitat and success rates, *Proceedings of the National Wild Turkey Symposium* 6:18–24.

Settle, R.H., G.A. Sommerville, J. McCormick, and D.M. Broom, 1994, Human scent matching using specially trained dogs, *Animal Behaviour* 48:1443–1448.

Shaffer, T.L., A.L. Dahl, R.E. Reynolds, K.L. Baer, M.A. Johnson, and G.A. Sargeant, 2006, Determinants on mallard and gadwall nesting on constructed islands in North Dakota, *Journal of Wildlife Management* 70:129–137.

Shallenberger, R.J., 1975, Olfactory use in the wedge-tailed shearwater (*Puffinus pacificus*) on Manana Is. Hawaii, in D.A. Denton and J.P. Coghlan, Eds., *Olfaction and Taste*, Academic Press, New York, pp. 355–359.

Shaw, R.H., G. den Hartog, and H.H. Neumann, 1988, Influence of foliar density and thermal stability on profiles of Reynolds stress and turbulence intensity in a deciduous forest, *Boundary-Layer Meteorology* 45:391–409.

Shivik, J.A., 2002, Odor-adsorptive clothing, environmental factors, and search-dog ability, *Wildlife Society Bulletin* 30:721–727.

Shuttleworth, W.J., 1989, Micrometeorology of temperature and tropical forest, *Philosophical Transactions of the Royal Society of London B* 324:299–334.

Sieving, K.E. and M.F. Willson, 1998, Nest predation and avian species diversity in northwestern forest understory, *Ecology* 79:2391–2402.

Singer, A.G., G.K. Beauchamp, and K. Yamazaki, 1997, Volatile signals of the major histocompatibility complex in male mouse urine, *Proceedings of the National Academy of Sciences* 94:2210–2214.

Skeel, M.A., 1983, Nesting success, density, philopatry, and nest-site selection on the whimbrel (*Numenius phaeopus*) in different habitats, *Canadian Journal of Zoology* 61:218–225.

Skogland, T., 1989, Comparative social organization of wild reindeer in relation to food, mates and predator avoidance, *Advances in Ethology* 29:1–74.

Slater, S.C., D. Rollins, R.C. Dowler, and C.B. Scott, 2001, *Opuntia*: a prickly paradigm for quail management in west-central Texas, *Wildlife Society Bulletin* 29:706–712.

Slovkin, J.M., 1982, Habitat requirements and evaluations, in J.W. Thomas and D.E. Toweill, Eds., *Elk of North America*, Stackpole Books, Harrisburg, PA, pp. 369–413.

Slovkin, J.M., P. Zager, and B.K. Johnson, 2002, Elk habitat selection and evaluation, in D.E. Toweill and J.W. Thomas, Eds., *North American Elk: Ecology and Management*, Smithsonian Institution Press, Washington, DC, pp. 531–556.

Smedman-Hogstrom, A.S. and U. Hogstrom, 1978, A practical method for determining wind frequency distributions for the lowest 200 m from routine meteorological data, *Journal of Applied Meteorology* 17:942–954.

Smith, D.A., K. Ralls, A. Hurt, B. Adams, M. Parker, B. Davenport, M.C. Smith, and J.E. Maldonado, 2003, Detection and accuracy rates of dogs trained to find scats of San Joaquin kit foxes (*Vuples macrotis mutica*), *Animal Conservation* 6:339–346.

Smith, S. A. and R.A. Paselk, 1986, Olfactory sensitivity of the turkey vulture (*Cathartes aura*) to three carrion-associated odorants, *Auk* 103:586–592.

Soane, I.D. and B. Clarke, 1973, Evidence for apostatic selection by predators using olfactory cues, *Nature* 241:62–64.

Soderstrom, B., T. Part, and J. Ryden, 1998, Different nest predator faunas and nest predation risk on ground and shrub nests at forest ecotones: an experiment and a review, *Oecologia* 117:108–118.

Sommerville, B. and D. Gee, 1984, Research on body odours: new prospects for combating crime, *International Criminal Police Review* 407(4):18–22.

Sonerud, G.A., 1985, Nest hole shift in Tengmalm's owl *Aegolius funereus* as defence against nest predation involving long-term memory in the predator, *Journal of Animal Ecology* 54:179–192.

Sonerud, G.A., 1989, Reduced predation by pine martens on nests of Tengmalm's owl in relocated boxes, *Animal Behaviour* 37:332–343.

Speake, D.W., T.E. Lynch, W.J. Fleming, G.A. Wright, and W.J. Hamrick, 1975, Habitat use and seasonal movements of wild turkeys in the Southeast, *Proceedings of the National Wild Turkey Symposium* 3:122–130.

Stacey, G.R., R.E. Belcher, C.J. Wood, and B.A. Gardiner, 1994, Wind flows and forces in a model spruce forest, *Boundary-Layer Meteorology* 69:311–334.

Stager, K.E., 1967, Avian olfaction, *American Zoologist* 7:415–420.

Steen, J.B. and E. Wilsson, 1986, A sense of direction on the scent of trouble, *New Scientist*, March 13:26.

Steen, J.B. and E. Wilsson, 1990, How do dogs determine the direction of tracks? *Acta Physiologica Scandinavica* 139:531–534.

Steigers, W.D., Jr. and J.T. Flinders, 1980, Mortality and movements of mule deer fawns in Washington, *Journal of Wildlife Management* 44:381–388.

Stoddard, D.M., 1980, *The Ecology of Vertebrate Olfaction*, Chapman and Hall, London.

Storaas, T. and P. Wegge, 1987, Nesting habitats and nest predation in sympatric populations of capercaillie and black grouse, *Journal of Wildlife Management* 51:167–172.

Stull, R.B., 1988, *An Introduction to Boundary Layer Meteorology*, Kluwer Academic Publishers, Dordrecht, The Netherlands.

Sugden, L.G. and G.W. Beyersbergen, 1987, Effect of nesting cover density on American crow predation of simulated duck nests, *Journal of Wildlife Management* 51:481–485.

Sullivan, T.P., 1986, Influence of wolverine (*Gulo gulo*) odor on feeding behavior of snowshoe hares (*Lepus americanus*), *Journal of Mammalogy* 67:385–388.

Sullivan, T.P. and D.R. Crump, 1986, Feeding responses of showshoe hares (*Lepus americanus*) to volatile constituents of red fox (*Vulpes vulpes*) urine, *Journal of Chemical Ecology* 12:729–739.

Sullivan, T.P., L.O. Nordstrom, and D.S. Sullivan, 1985, Use of predator odors as repellents to reduce feeding damage by herbivores. I. Showshoe hares (*Lepus americanus*), *Journal of Chemical Ecology* 11:903–919.

Sullivan, T.P., D.S. Sullivan, D.R. Crump, H. Weiser, and E.A. Dixon, 1988, Predator odors and their potential role in managing pest rodents and rabbits, *Vertebrate Pest Conference* 13:145–150.

Sun, J. and L. Mahrt, 1995, Relationship of surface heat flux to microscale temperature variations: applications to BOREAS, *Boundary-Layer Meteorology* 76:291–301.

Sutton, O.G., 1953, *Micrometeorology,* McGraw-Hill Book, New York.

Sveum, C.M., W.D. Edge, and J.A. Crawford, 1998, Nesting habitat selection by sage grouse in south-central Washington, *Journal of Range Management* 51:265–269.

Swanson, D.A., J.C. Pack, C.I. Taylor, D.E. Samuel, and P.W. Brown, 1996, Selective timber harvesting and wild turkey reproduction in West Virginia, *Proceedings of the National Wild Turkey Symposium* 7:81–88.

Sweeney, J.R., R.L. Marchinton, and J.M. Sweeney, 1971, Responses of radio-monitored white-tailed deer chased by hunting dogs, *Journal of Wildlife Management* 35:707–716.

Swennen, C., 1968, Nest protection of eiderducks and shovelers by means of faeces, *Ardea* 56:248–258.

Syrotuck, W.G., 1972, *Scent and the Scenting Dog*, Arner Publications, Canastota, NY.

Tampieri, F., 1987, Separation features of boundary-layer flow over valleys, *Boundary-Layer Meteorology* 40:295–307.

Taylor, J.S., K.E. Church, and D.H. Rusch, 1999, Microhabitat selection by nesting and brood-rearing northern bobwhite in Kansas, *Journal of Wildlife Management* 63:686–694.

Taylor, P.A. and R.J. Lee, 1984, Simple guidelines for estimating wind speed variations due to small scale topographic features, *Climatological Bulletin* 18(2):3–42.

Taylor, P.A., P.J. Mason, and E.F. Bradley, 1987, Boundary-layer flow over low hills, *Boundary-Layer Meteorology* 39:107–132.

Tegt, J.L., 2004, Coyote (*Canis latrans*) Recognition of Relatedness Using Odor Cues in Feces, Urine, Serum and Anal Sac Secretions, M.S. thesis, Utah State University, Logan.

Thogmartin, W.E., 1999, Landscape attributes and nest-site selection in wild turkeys, *Auk* 116:912–923.

Thom, A.S., J.B. Stewart, H.R. Oliver, and J.H.C. Gash, 1975, Comparison of aerodynamic and energy budget estimates of fluxes over a pine forest, *Quarterly Journal of the Royal Meteorological Society* 101:93–105.

Thompson, O.E. and R.T. Pinker, 1975, Wind and temperature profile characteristics in a tropical evergreen forest in Thailand, *Tellus* 27:562–573.

Tirpak, J.M., W.M. Giuliano, C.A. Miller, S. Bittner, J.W. Edwards, S. Friedhof, W.K. Igo, D.F. Stauffer, and G.W. Norman, 2006, Ruffed grouse nest success and habitat selection in the central and southern Appalachians, *Journal of Wildlife Management* 70:138–144.

Townsend, G.H., 1966, A study of waterfowl nesting on the Saskatchewan River Delta, *Canadian Field-Naturalist* 80:74–88.

Tremblay, J.P., G. Gauthier, D. Lepage, and A. Desrochers, 1997, Factors affecting nesting success in greater snow geese: effects of habitat association with snowy owls, *Wilson Bulletin* 109:449–461.

Turner, D.B., 1967, *Workbook of Atmospheric Dispersion Estimates*, U.S. Department of Heath, Education and Welfare, National Center for Air Pollution Control, Cincinnati, OH.

Uresk, D.W., T.A. Benzon, K.E. Severson, and L. Benkobi, 1999, Characteristics of white-tailed deer fawn beds, Black Hills, South Dakota, *Great Basin Naturalist* 59:348–354.

Vale, G.A., 1977, The flight of tsetse flies (Diptera: Glossinidae) to and from a stationary ox, *Bulletin of Entomological Research* 67:297–303.

Van Balen, J.H., C.J.H. Booy, J.A. van Franeker, and E.R. Osieck, 1982, Studies on hole-nesting birds in natural nest sites. 1. Availability and occupation of natural nest sites, *Ardea* 70:1–24.

Voskamp, K.E., C.J. Den Otter, and N. Noorman, 1998, Electroantennogram responses of tsetse flies (*Glossina pallidipes*) to host odours in an open field and riverine woodland, *Physiological Entomology* 23:176–183.

Waldrip, G.P. and J.H. Shaw, 1979, Movements and habitat use by cow and calk elk at the Wichita Mountains National Wildlife Refuge, in M.S. Boyce and L.D. Hayden-Wing, Eds., *North American Elk: Ecology, Behavior and Management*, University of Wyoming Press, Laramie, pp. 177–184.

Walker, J.C. and R.A. Jennings, 1991, Comparison of odor perception in humans and animals, in D.G. Laing, R.L. Doty, and W. Breipohl, Eds., *The Human Sense of Smell*, Springer-Verlag, New York, pp. 261–280.

Wallace, M.C. and P.R. Krausman, 1992, Neonatal elk habitat in central Arizona, in R.D. Brown, Ed., *The Biology of Deer*, Spinger-Verlag, New York, pp. 69–75.

Ward, J.M. and P.L. Kennedy, 1996, Effects of supplemental food on size and survival of juvenile northern goshawks, *Auk* 113:200–208.

Wark, K. and C.F. Warner, 1976, *Air Pollution: Its Origin and Control*, Harper and Row, New York.

Wark, K. and C.F. Warner, 1981, *Air Pollution: Its Origin and Control*, 2nd ed., Harper and Row, New York.

Welty, J.C., 1962, *The Life of Birds*, W.B. Saunders, Philadelphia.

Whelan, C.J., M.L. Dilger, D. Robson, N. Hallyn, and S. Dilger, 1994, Effects of olfactory cues on artificial-nest experiments, *Auk* 111:945–952.

White, C.G., S.H. Schweitzer, C.T. Moore, I.B. Parnell III, and L.A. Lewis-Weis, 2005, Evaluation of the landscape surrounding northern bobwhite nest sites: a multiscale analysis, *Journal of Wildlife Management* 69:1528–1537.

White, M., F.F. Knowlton, and W.C. Glazener, 1972, Effects of dam-newborn fawn behavior on capture and mortality, *Journal of Wildlife Management* 36:897–906.

Whiteman, C.D., 1982, Breakup of temperature inversions in deep mountain valleys: part 1. Observations, *Journal of Applied Meteorology* 21:270–289.

Whitten, W.K., M.C. Wilson, S.R. Wilson, J.W. Jorgenson, M. Novotny, and M. Carmack, 1980, Induction of marking behavior in wild red foxes (*Vulpes vulpes* L.) by synthetic urinary constituents, *Journal of Chemical Ecology* 6:49–55.

Wiebe, K.L., 2001, Microclimate of tree cavity nests: is it important for reproductive success in northern flickers? *Auk* 118:412–421.

Wiebe, K.L. and K. Martin, 1998, Costs and benefits of nest cover for ptarmigan: changes within and between years, *Animal Behaviour* 56:1137–1144.

Wiebe, K.L. and T. L. Swift, 2001, Clutch size relative to tree cavity size in northern flickers, *Journal of Avian Biology* 32:167–173.

Wilcove, D.S., 1985, Nest predation in forest tracts and the decline of migratory songbirds, *Ecology* 66:1211–1214.

Wilcove, D.S., C.H. McLellan, and A.P. Dobson, 1986, Habitat fragmentation in the temperate zone, in M.E. Soule, Ed., *Conservation Biology: The Science of Scarcity and Diversity*, Sinauer Associates, Sunderland, MA, pp. 237–256.

Willis, M.A. and T.C. Baker, 1984, Effects of intermittent and continuous pheromone stimulation on the flight behaviour of the oriental fruit moth, *Grapholita molesta, Physiological Entomology* 9:341–358.

Willis, M.A. J. Murlis, and R.T. Carde, 1991, Pheromone-mediated upwind flight of male gypsy moths, *Lymantria dispar*, in a forest, *Physiological Entomology* 16:507–521.

Winter, M., D.H. Johnson, and J. Faaborg, 2000, Evidence for edge effects on multiple levels in tallgrass prairie, *Condor* 102:256–266.

Winter, M., D.H. Johnson, and J.A. Shaffer, 2005, Variability in vegetation effects on density and nest success of grassland birds, *Journal of Wildlife Management* 69:185–197.

With, K.A., 1994, The hazards of nesting near shrubs for a grassland bird, the McCown's longspur, *Condor* 96:1009–1019.

With, K.A. and D.R. Webb, 1993, Microclimate of ground nests: the relative importance of radiative cover and wind breaks from three grassland species, *Condor* 95:401–413.

Wolff, J.O., T. Fox, R.R. Skillen, and G. Wang, 1999, The effects of supplemental perch sites on avian predation and demography of vole populations, *Canadian Journal of Zoology* 77:535–541.

Wright, R.H., 1964, *The Science of Smell*, Basic Books, New York.

Xian, X., W. Tao, S. Qingwei, and Z. Weimin, 2002, Field and wind-tunnel studies of aerodynamic roughness length, *Boundary-Layer Meteorology* 104:151–163.

Yahner, R.H. and B.L. Cypher, 1987, Effects of nest location on depredation of artificial arboreal nests, *Journal of Wildlife Management* 51:178–181.

Yan, H. and R.A. Anthes, 1987, The effect of latitude on the sea breeze, *Monthly Weather Review* 115:936–956.

Yoshino, M.M., 1975, *Climate in a Small Area: An Introduction to Local Meteorology*, University of Tokyo Press, Japan.

Appendix one

Latin names of species mentioned in this book

Albatross, Black-footed—*Phoebastria nigripes*
Albatross, Wandering—*Diomedea exulans*
Aspen—*Populus tremuloides*
Avocet, American—*Recurvirostra americana*
Badger, American—*Taxidea taxus*
Bear, Black—*Ursus americanus*
Bear, Grizzly—*Ursus arctos*
Beaver—*Castor canadensis*
Bentgrass, Colonial—*Agrostis tenuis*
Bison, American—*Bison bison*
Blackbuck, Indian—*Antilope cervicapra*
Blackstart—*Cercomela melaura*
Bluegrass, Kentucky—*Poa pratensis*
Bluestem, Little—*Schizachyrium scoparium*
Bluestem, Sand—*Andropogon hallii*
Bobcat—*Lynx rufus*
Bobolink—*Dolichonyx oryzivorus*
Bobwhite, Northern—*Colinus virginianus*
Brome, Smooth—*Bromus* spp.
Cactus, Prickly pear—*Opuntia* spp.
Cactus, Saguaro—*Camegiea gigantes*
Canarygrass, Reed—*Phalaris arundinacea*
Capercaillie—*Tetrao urogallus*
Caribou—*Rangifer tarandus*
Chipmunk—*Tamias* spp. or *Eutamias* spp.
Chipmunk, Eastern—*Tamias striatus*
Cormorant, Double-crested—*Phalacrocorax auritus*
Cougar—*Felis concolor*
Coyote—*Canis latrans*
Crow, American—*Corvus brachyrhynchos*
Deer, Axis—*Axis axis*

Deer, Fallow—*Dama dama*
Deer, Mule—*Odocoileus hemionus*
Deer, Red—*Cervus elaphus*
Deer, Roe—*Capreolus capreolus*
Deer, White-tailed—*Odocoileus virginianus*
Dickcissel—*Spiza americana*
Dik-dik—*Madoqua* spp.
Dog, African wild—*Lycaon pictus*
Dog, Domestic—*Canis familiaris*
Dog, Prairie—*Cynomys* spp.
Dove, Mourning—*Zenaida macroura*
Duck, Ring-necked—*Aythya collaris*
Eagle, Golden—*Aquila chrysaetos*
Eider, Common—*Somateria mollissima*
Elk—*Cervus canadensis*
Fern, Bracken—*Pteridium aquilinum*
Fisher—*Martes pennanti*
Flicker—*Colaptes* spp.
Flicker, Northern—*Colaptes auratus*
Fly, Oriental Fruit—*Grapholita molesta*
Fly, Tsetse—*Glossina* spp.
Flycatcher, Collared—*Ficedula albicolliss*
Flycatcher, Dusky—*Empidonax oberholseri*
Flycatcher, Pied—*Ficedula hypoleuca*
Flycatcher, Willow—*Empidonax traillii*
Fox, Arctic—*Alopex lagopus*
Fox, Kit—*Vulpes macrotis*
Fox, Red—*Vulpes vulpes*
Gadwall—*Anas strepera*
Gazelle—*Gazella* spp.
Gazelle, Dorcas—*Gazella dorcas*
Gazelle, Grant's—*Gazella granti*
Gazelle, Thomson's—*Gazella thomsoni*
Giant-petrel—*Macronectes* spp.
Goat, Feral—*Capra hircus*
Goose, Canada—*Branta canadensis*
Goose, Snow—*Chen caerulescens*
Goshawk, Northern—*Accipiter gentilis*
Grackle, Common—*Quiscalus quiscula*
Grass, Grama—*Bouteloua* spp.
Grass, Indian—*Sorghastrum nutans*
Grosbeak, Evening—*Coccothraustes vespertinus*
Grouse, Black—*Tetrao tetrix*
Grouse, Ruffed—*Bonasa umbellus*
Grouse, Sharp-tailed—*Tympanuchus phasianellus*
Gull, California—*Larus californicus*
Gull, Ring-billed—*Larus delawarenis*
Hare, Snowshoe—*Lepus americanus*
Hawk, Ferruginous—*Buteo regalis*
Hog, Feral—*Susscrofa*
Horse—*Equus caballus*

Hyena, Brown—*Hyaena brunnea*
Hyena, Spotted—*Crocuta crocuta*
Ibex, Siberian—*Capra ibex siberica*
Jackrabbit—*Lepus* spp.
Jay, Blue—*Cyanocitta cristata*
Jay, Gray—*Perisoreus canadensis*
Jay, Pinyon—*Gymnorhinus cyanocephalus*
Junco, Dark-eyed—*Junco hyemalis*
Kestrel, American—*Falco sparverius*
Klipspringer—*Oreotragus oreotragus*
Kob, Uganda—*Adenota kob thomasi*
Kudu—*Tragelaphus* spp.
Kudu, Greater—*Tragelaphus strepsiceros*
Kudu, Lesser—*Tragelaphus imberbis*
Lark, Horned—*Eremophila alpestris*
Lechwe, Red—*Kobus leche*
Longspur, McCown's—*Calcarius mccownii*
Lynx—*Lynx canadensis*
Macaque, Bonnet—*Macaca radiata*
Macaque, Stump-tailed—*Macaca arctoides*
Magpie, Black-billed—*Pica pica*
Mallard—*Anas platyrhynchos*
Marmot, Yellow-bellied—*Marmota flaviventris*
Marten, American Pine—*Martes americana*
Marten, European Pine—*Martes martes*
Marten, Pine—*Martes* spp.
Meadowlark, Eastern—*Sturnella magna*
Meadowlark, Western—*Sturnella neglecta*
Mink—*Mustela vison*
Monkey, Talapoin—*Miopithecus talapoin*
Moose—*Alces alces*
Moth, Gypsy—*Lymantria dispar*
Moth, Light-brown apple—*Epiphyas postvittana*
Moth, Pea—*Cydia nigricana*
Muntjac—*Muntiacus muntjak*
Muskox—*Ovibos moschatus*
Nettle, Stinging—*Urtica dioica*
Opossum—*Didelphis virginiana*
Ovenbird—*Seiurus aurocapilla*
Owl, Elf—*Micrathene whitneyi*
Owl, Tengmalm's—*Aegolius funereus*
Palmetto, Saw—*Serenoa repens*
Palo Verde—*Circidium* spp.
Partridge, Gray—*Perdix perdix*
Partridge, Red-legged—*Alectoris rufa*
Petrel, Cape—*Daption capensis*
Petrel, Leach's—*Oceanodroma leucorrhoa*
Petrel, White-chinned—*Procellaria aequinoctialis*
Pheasant, Ring-neck—*Phasianus colchicus*
Pika—*Ochotona princeps*
Pine, Jack—*Pinus banksiana*

Pintail, Northern—*Anas acuta*
Plover, Mountain—*Charadrius montanus*
Poorwill, Common—*Phalaenoptilus nuttallii*
Prairie-chicken, Greater—*Tympanuchus cupido*
Prairie-chicken, Lesser—*Tympanuchus pallidicinctus*
Pronghorn—*Antilocapra americana*
Ptarmigan, Svalbard—*Lagopus mutus*
Ptarmigan, White-tailed—*Lagopus leucura*
Ptarmigan, Willow—*Lagopus lagopus*
Pudu—*Pudu pudu*
Raccoon—*Procyon lotor*
Reedbuck—*Redunca* spp.
Robin, American—*Turdus migratorius*
Sagebrush—*Artemisia* spp.
Sagebrush, Big—*Artemisia tridentata*
Sage-grouse, Greater—*Centrocercus urophasianus*
Sapsucker—*Sphyrapicus* spp.
Scaup, Lesser—*Aythya affinis*
Shearwater, Wedged-tailed—*Puffinus pacificus*
Sheep, Barbary—*Ammotragus lervia*
Sheep, Bighorn—*Ovis canadensis*
Sheep, Dall's—*Ovis dalli*
Sheep, Mouflon—*Ovis musimon*
Sitatunga—*Tragelaphus spekii*
Skunk, Hognosed—*Conepatus leuconotus*
Skunk, Striped—*Mephitis mephitis*
Snake, Brown—*Storeria dekayi*
Snake, Garter—*Thamnophis* spp.
Snake, Northern Water—*Nerodia sipedon*
Sparrow, Clay-colored—*Spizella pallida*
Sparrow, Field—*Spizella pusilla*
Sparrow, Grasshopper—*Ammodramus savannarum*
Sparrow, Harris—*Zonotrichia querula*
Sparrow, Savannah—*Passerculus sandwichensis*
Sparrow, Vesper—*Pooecetes gramineus*
Sparrow, White-crowned—*Zonotrichia leucophrys*
Springbok—*Antidorcas marsupialis*
Spruce, Black—*Picea mariana*
Spruce, Sitka—*Sicea sitchensis*
Squirrel, Douglas—*Tamiasciurus douglasii*
Squirrel, Franklin's ground—*Spermophilus franklinii*
Squirrel, Red—*Tamiasciurus hudsonicus*
Squirrel, Southern flying—*Glaucomys volans*
Starling, European—*Sturnus vulgaris*
Stoat—*Mustela erminea*
Storm-Petrel, Black-bellied—*Fregetta tropica*
Storm-Petral, Leach's—*Oceanodroma leucorrhoa*
Storm-Petrel, Wilson's—*Oceanites oceanicus*
Swan, Mute—*Cygnus olor*
Tanager, Scarlet—*Piranga olivacea*
Tern, Forster's—*Sterna forsteri*

Tern, Sandwich—*Thalasseus sandvicensi*
Thrush, Hermit—*Catharus guttatus*
Thrush, Song—*Turdus philomelos*
Thrush, Wood—*Hylocichla mustelina*
Tit, Great—*Parus major*
Tsetse Fly—*Glossina* spp.
Verdin—*Auriparus flaviceps*
Vireo, Red-Eyed—*Vireo olivaceus*
Vole, Montane—*Microtus montanus*
Vulture, Turkey—*Cathartes aura*
Warbler, MacGillivray's—*Oporornis tolmiei*
Warbler, Yellow—*Dendroica petechia*
Waterbuck, Defassa—*Kobus ellipsiprymnus defassa*
Weasel, Long-Tailed—*Mustela frenata*
Weasel, Short-Tailed—*Mustela erminea*
Weaver, Baya—*Ploceus philippinus*
Whimbrel—*Numenius phaeopus*
Wildebeest—*Connochaetes taurinus*
Wolf—*Canis lupus*
Wolverine—*Gulo gulo*
Woodpecker, Black—*Cryocpus martius*
Woodpecker, Gila—*Melanerpes uropygialis*
Woodpecker, Pileated—*Dryocopus pileatus*
Wren, Cactus—*Campylorhynchus brunneicapillus*

Appendix two

Symbols used in this book

cm   Centimeter .

d   Zero-plane displacement. When plants or other surfaces features are packed close enough together, the air skims over them so that they act as displaced surface to the wind. The region between this displaced surface and the actual ground surface is d.

g   Gram.

h   Height of surface features, such as a forest's canopy.

i   Turbulence intensity: $I = [(u^2 + v^2 + w^2)/U]^{1/2}$.

K   Potential temperature (measured relative to absolute zero or $-273°C$).

kg   Kilogram.

km   Kilometer.

m   Meter.

mm   Millimeter.

sec   Second.

u   Velocity of wind that is moving horizontally and in the stream-wise direction.

u_*   Friction velocity.

u_*/U   Bulk drag coefficient.

U   Mean wind velocity in the stream-wise direction during a period of time.

$u'w'$ Shear stress.

sec   Second.

v   Lateral wind velocity. This is airflow that is horizontal and perpendicular to U (i.e., along the v axis).

V   Mean lateral wind velocity during a period of time.

x   Distance downwind (leeward) of an object.

$-x$ Distance upwind (windward) of an object.

w   Vertical wind velocity (along the w axis).

W   Mean vertical wind velocity during a period of time.

z   Height above the ground.

z/h   Height relative to a surface feature (a value of 0.5 would mean a height halfway up the surface feature).

z_0   Aerodynamic roughness length. This is the height above the ground where $U = 0$ and eddies begin to predominate.

Δ   Change in something.

ΔS Change in wind velocity at the top of a surface feature relative to the wind speed measured elsewhere in the boundary layer: $\Delta S = (U_{(\text{at top of hill})} - U_{(\text{normal})}) / U_{(\text{normal})}$.

σ Standard deviation.

σ_u stream-wise turbulence (standard deviation of the wind velocity in the stream-wise direction or along the u axis).

σ_u/U stream-wise turbulence intensity (stream-wise turbulence relative to mean wind velocity).

σ_v Lateral turbulence (standard deviation of the lateral wind velocity or along the v axis).

σ_v/U Lateral turbulence intensity (lateral turbulence relative to mean wind velocity).

σ_w Vertical turbulence (standard deviation of the vertical wind velocity or along the w axis).

σ_w/U Vertical turbulence intensity (vertical turbulence relative to mean wind velocity).

Λ Canopy area index. This is a simplified version of λ. It is the single-sided area of all surface elements in the canopy (leaves, stems, etc.) divided by the ground area beneath them.

λ Frontal area index of surface features. This is the ratio of a surface area that is perpendicular to the wind divided by amount of ground area the surface feature occupies (λ = frontal surface area / ground area occupied).

Appendix three

Forces controlling wind speed and direction

The differential heating of the atmosphere produces pressure gradients that vary across the face of the earth. It is this variation in atmospheric pressure that produces large-scale motions of the air. Atmospheric pressure is measured using barometers. High-pressure areas are usually colder areas; low-pressure areas are usually warmer. These areas are designated on weather maps by connecting areas of equal pressure with lines called *isobars* (Figure A3.1). The greatest change in air pressure occurs when there is a great difference in atmospheric pressure between a high-pressure and a low-pressure center and the two centers are near to each other. This can be seen on weather maps by a compression of the isobars, and it is in these areas where strong winds occur.

Air flows from areas of high pressure to low pressure. If the earth were not spinning, then the wind would flow directly between the two (e.g., perpendicular to the isobars). However, the earth's rotation alters this situation by creating another force called the *Coriolis force*. This force is produced because an air mass at the equator rotates around the earth's axis at the same speed as the ground. Because the earth's circumference at the

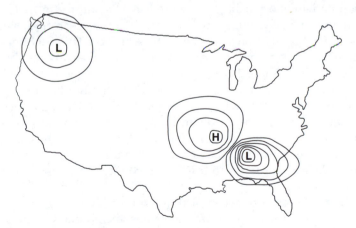

Figure A3.1 Weather map for the United States that shows isobars around high- and low-pressure areas.

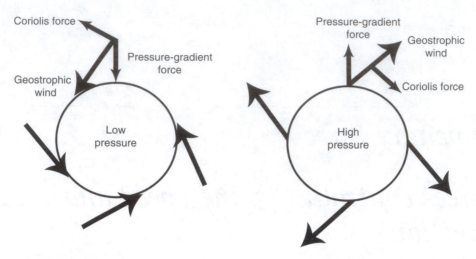

Figure A3.2 Vectors showing how the pressure gradient force and the Coriolis force combine to create the geostrophic wind.

equator is larger than anywhere else on the planet, an equatorial air mass is moving toward the east faster than the surface of the earth is moving at higher latitudes. Because of this, if an equatorial air mass starts moving directly north toward the North Pole, it will veer eastward because its centrifugal speed is now faster than the earth's surface below it. In the Southern Hemisphere, the Coriolis force causes air masses to veer to the west.

The Coriolis force has an impact not only on equatorial air masses, but also on any air mass that is changing latitudes. The Coriolis force is always perpendicular to the wind's direction. The wind that results from the combination of the pressure gradient force and the Coriolis force is called the *geostrophic wind* (Figure A3.2). Because of the Coriolis force, the geostrophic wind does not move directly from high-pressure to low-pressure areas. Instead, it moves clockwise around a high-pressure area and counterclockwise around a low-pressure area in the Northern Hemisphere (Figure A3.2) and in the opposite directions in the Southern Hemisphere.

The geostrophic winds occur high above the earth's surface in the free atmosphere. Near the ground, the surface wind is much slower than the geostrophic wind because the surface wind is slowed by friction with the earth's surface. This frictional force is in the opposite direction of the wind and increases with both decreasing height and increasing surface roughness (Figure A3.3). The combination of the frictional force, pressure force, and Coriolis force also causes the surface wind to turn slightly in a counterclockwise direction. This causes the surface winds to be deflected toward the direction of the low-pressure area. Hence, the wind direction at the ground may not be a good prediction of wind direction further aloft. This shift in wind direction close to the earth is often minor when surface wind speeds are high (i.e., greater than 5 m/second), but when winds are calm, the angle of deflection can be 15 to 45° in a counterclockwise direction (Wark and Warner 1976).

One interesting difference between airflow around high- and low-pressure areas is that, because the air flows toward low-pressure areas, the air mass in the center must rise. In doing so, it carries moisture aloft, where it cools, condenses, and often falls as rain or snow. This is why low-pressure areas are often associated with precipitation. Air flows away from high-pressure areas, and this causes air from high altitudes to sink. This high-altitude air warms as it sinks. High-altitude air contains little moisture, and its capacity

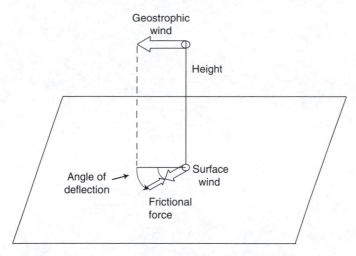

Figure A3.3 Vectors showing the geostrophic and surface winds (the length of the vectors corresponds to the wind speed). The surface wind is slower than geostrophic wind because it is opposed by frictional force and its direction has been deflected counterclockwise.

to hold water increases as it sinks and warms. The end result is that high-pressure areas usually produce sunny skies.

Appendix four

Pasquill's system for measuring atmospheric stability

Pasquill (1961) developed a classification system for atmospheric stability consisting of six categories that were designated by the letters A–F (Table A4.1). One advantage of the Pasquill system is its simplicity; people can classify local atmospheric conditions into one of his categories based on wind speed and insolation (Table A4.2), with insolation estimated by the angle of the sun and the thickness of the cloud cover (Table A4.3). Another advantage of Pasquill's system is that it provides reasonable estimates for the values σ_v and σ_w used in the Gaussian dispersion model. Because σ_v and σ_w vary with atmospheric stability, these values are highest in Stability Class A (strong thermals) and lowest in Stability Class F (stable conditions). The rates of dispersion along a lateral (σ_v) and vertical

Table A4.1 Pasquill's Atmospheric Stability Categories

Stability category	Classification	Atmospheric condition	Most likely to happen	Frequency of occurrence in England
A	Extremely unstable	Bright sun, strong thermal	Day, summer	1%
B	Moderately unstable	Moderate mixing, transitional periods	Day, all year	6%
C	Slightly unstable	Slight mixing, transitional periods	Day, all year	17%
D	Neutral	Strong winds, overcast	Day or night, all year	60%
E	Slightly stable	Moderate winds, transitional periods	Night, all year	7%
F	Moderately stable	Clear night skies, light winds	Night, all year	9%

Source: Adapted from Barratt, R., *Atmospheric Dispersion Modelling: An Introduction to Practical Applications*, Earthscan Publications, Sterling, VA, 2001, and used with permission of Earthscan Publications).

Table A4.2 Wind Velocity and Radiation Levels (Solar Radiation during the Day[a] and Black Body Radiation at Night[b]), Characteristic of Different Pasquill Stability Categories

Wind velocity (m/second)[c]	Day (solar radiation)			Night (black-body radiation)	
	Strong	Moderate	Slight	Sky > 50% overcast	Sky < 50% overcast
0–2	A	A–B	B	F	F
2–3	A–B	B	C	E	F
3–5	B	B–C	C	D[d]	D
5–6	C	C–D	D	D	D
>6	C	D	D	D	D

[a] See Table A4.3 for further refinement of solar radiation levels.

[b] Night extends from 1 h before sunset to 1 h after sunrise.

[c] Wind velocity is measured at the standard height of 10 m above the ground.

[d] Pasquill's Stability Category D exists for all overcast conditions, day or night, regardless of wind velocity.

Source: Adapted from Barratt, R., *Atmospheric Dispersion Modelling: An Introduction to Practical Applications*, Earthscan Publications, Sterling, VA, 2001, and used with permission of Earthscan Publications.

Table A4.3 Solar Radiation Levels Based on the Angle of the Sun and Cloud Cover

Cloud cover	Angle of the sun (elevation)		
	<35°	35–60°	>60°
<50% low cloud cover or only high, thin clouds	Slight	Moderate	Strong
50–85% midlevel clouds	Slight	Slight	Moderate
>60% low-level clouds or >85 midlevel clouds	Slight	Slight	Slight

Source: Adapted from Barratt, R., *Atmospheric Dispersion Modelling: An Introduction to Practical Applications*, Earthscan Publications, Sterling, VA, 2001, and used with permission of Earthscan Publications.

(σ_w) direction can vary tenfold depending on the stability of the atmosphere (Figures 3.11 through 3.13). Although not shown in these figures and tables, values for σ_u are similar to σ_v. A high value for σ_u indicates that the wind is gusty and often changes speed. Such conditions cause volatility in odorant concentration directly downwind of the odorant source.

Author Index

A

Albone, E. S., 11
Albrecht, T., 173
Alcock, J., 6
Aldeman, S., 10
Aldridge, C. L., 187
Alldredge, A. W., 151
Amiro, B. D., 114
Anderson, J. R., 69
Andren, H., 188, 192, 196
Angelstam, P., 188, 192
Arnold, T. W., 173
Askins, R. A., 192
Atema, J., 57
Austin, G. T., 164, 166
Autenrieth, R. E., 146, 148
Axel, R., 3
Aylor, D. E., 111

B

Badyaev, A. V., 129, 133
Baines, D., 188
Baldocchi, D. D., 97, 110, 114, 189
Balgooyen, T. C., 166
Ballard, W. B., 152
Barka, T., 10
Barratt, R., 49, 50, 98
Barrett, M. W., 96, 151
Barrette, C., 10
Batary, P., 163, 192
Batchelor, R. F., 154
Bednekoff, P. A., 137, 138
Bekesy, G., 57
Bekoff, M., 131, 142, 176
Bergen, J. D., 118
Berger, J., 156, 157
Bergerud, A. T., 144, 155, 156, 176
Best, L. B., 159, 160, 170, 185
Blackadar, A. K., 97
Bleich, V. C., 156, 157
Bollinger, E. K., 6, 179, 188
Borgo, J. S., 12

Bossert, W. H., 41, 42, 43
Bowman, G. B., 170, 183, 184
Bowyer, R. T., 144
Bradley, E. F., 79, 99
Bradley, R. M., 10
Brady, J., 56, 59, 60, 61, 118, 119
Brighton, P. W. M., 80
Brittingham, M. C., 196
Brua, R. B., 173
Budgett, H. M., 13, 24, 27, 28, 29, 32, 37, 172
Bump, G., 176
Burghardt, G. M., 6
Burtt, E. H., 11
Byers, S. M., 183

C

Cain, J. W., 129
Calladine, J., 188
Cannon, S. K., 152
Carbone, C., 123
Carde, R. T., 47, 116
Caro, T., 4, 6, 12, 13, 129, 130, 137, 144, 147, 160, 161, 162, 163, 168, 169, 177
Carroll, B. K., 4, 140, 150
Caughey, S. J., 65
Chalfoun, A. D., 159, 192
Chesness, R. A., 133
Christensen, T. A., 57
Clapperton, B. K., 12
Clark, C. W., 138
Clark, L., 4, 37
Clark, R. G., 13, 169
Clifford, R. J., 27
Connelly, J. W., 124
Conner, R. N., 166
Conover, M. R., 14, 32, 127
Cook, R. S., 11, 139, 150, 152
Cowardin, L. M., 189
Crabtree, R. L., 183
Creel, S., 123
Crockett, A. B., 164
Czaplicki, J. A., 6

Species Index

Subject Index

A

Aerodynamic roughness length
    definition, 96–98
    effect on nest success, 181–185, 189–192, 202
    effect on olfactory predators, 98, 124, 181–185,
        189–192, 202
    symbol for, 96–98, 229
    values for different surfaces, 97
Airborne odorants. *See* odorant plume.
Artificial nest experiments, 14–15, 71–76
Atmosphere, layers of, 47–48. *See also* boundary layer,
        free atmosphere, *and* surface layer.
Atmospheric instability
    defined, 47–54
    dinural changes in, 50–53
    effect on olfactory predators, 53–54
    Pasquill's classification system, 235–236
Atmospheric stability. *See* atmospheric instability.

B

Bedding sites. *See* partuition.
Boundary layer, 48, 117, 232–233
Brownian motion, 19

C

Calves. *See* neonates.
Canopy area index
    definition, 100
    effect on aerodynamic roughness length, 100–101
    symbol, 230
    within grass canopies, 110–111
Cavity nests. *See* nests.
Convection. *See* updrafts *and* turbulence, vertical.
Coriolis force, 47–48, 231–233

D

Death feigning, 161
Dense nesting cover
    definition, 98

    effect on nest success, 169–170, 181–184
    effect on nesting ducks, 98, 181–184
    impeding movements of predators, 181–184
Depositional–odor trails
    ability of predators to follow, 29–30, 33–34, 201
    ability of prey to throw off predators, 30–33, 201
    creation, 23–24
    determining age, 24–25, 34
    determining direction, 25–26, 34
    impact of different surfaces on, 27–28
    impact of weather on, 27–28, 34
    structure, 23–24
Diffusion, 19–20
Diurnal
    behavior of neonates, 150
    effect on air movements within a plant's canopy,
        107–108, 120
    effect on atmospheric stability, 50–54, 235–236
    effect on depositional-odor trails, 28, 201
    effect on odorant plumes, 50–53, 201–202
    nocturnal winds and effect on nest success, 164,
        171–172
    trade-offs involving the timing of dangerous
        activities, 137–139
    why animals are active during corpuscular
        periods, 137–139
Drainage winds, 65
Dust devils, 51–53

E

Edge Habitats
    definitions, 196–198
    effect on aerodynamic roughness length, 101–104
    effect on forest porosity, 102–103
    effect on nest success, 192–194
    effect on olfactory predators, 104–105, 203
    effect on turbulence, 101–105
    effect on updrafts, 101–105
    effect on zero–plane displacement, 101–104
Efficient digester hypothesis, 138
Environmental lapse rate, 48
Evaporation, 16–17, 38–39